The Right to Knowledge

Julia Bello-Bravo • Anne Namatsi Lutomia •
John William Medendorp •
Barry Robert Pittendrigh

The Right to Knowledge

A Learning-Systems Approach for the Sustainable Development Goals

Springer

Julia Bello-Bravo
Department of Agricultural Sciences
Education and Communication
Purdue University
West Lafayette, IN, USA

John William Medendorp
Department of Entomology
Purdue University
West Lafayette, IN, USA

Anne Namatsi Lutomia
Department of Entomology
Purdue University
West Lafayette, IN, USA

Barry Robert Pittendrigh
Department of Entomology
Purdue University
West Lafayette, IN, USA

ISBN 978-3-032-03359-8 ISBN 978-3-032-03360-4 (eBook)
https://doi.org/10.1007/978-3-032-03360-4

This Springer imprint is published by the registered company Springer Nature Switzerland AG
The registered company address is: Gewerbestrasse 11, 6330 Cham, Switzerland

Preface

When we began writing this book, the international development sector was not yet in its present state of flux (possible free-fall). The dissolution of the United States Agency for International Development (USAID), which provided nearly $43.8 billion in aid in fiscal 2023, or about 60% of all foreign-assistance dollars (DeSilver, 2025), has instigated a vacuum in the international development system that is radically transforming it and will likely drive a major reassessment of the many assumptions and habits that maintained the system previously—especially a reexamination of its return-on-investment numbers in light of the massive outlays of resources.

Whatever rises from the ashes of the current deconstruction of the international development system, one point is certain: donors need to rethink their investments, which have often tolerated efforts to shore up the creaky structures of in-person knowledge dissemination. Costly investments in extension systems requiring an agent, transportation, lodging, and materials to reach an audience of dozens do not make sense moving forward. Flying experts regionally or across the globe to give decontextualized answers to an audience of hundreds at best—answers that will soon be forgotten because they meet the needs of no one in particular—are not cost-effective for donor investments. There needs to be a new push to wean systems of cost-ineffective approaches and redirect development efforts from conventional in-person platforms toward mass-scaled, cost-effective, and inclusive information and communication (ICT) platforms for the benefit of everyone.

For more than three decades now, the promise of ICT has been held out as the hope of international development. Nevertheless, that promise is rarely realized adequately when donors, governments, institutions, and organizations adhere too closely to outmoded, habitual, or inefficient practices. It is past time for a full transition to ICT knowledge dissemination to be effectively implemented. Without this transition, the dream of maximally reaching the Sustainable Development Goals will remain out of reach.

We had originally intended to open this book by highlighting the emergence of a new era—the post-World War II and post-Bretton Woods *Information or Digital Age*. But now, even as we write, a newer age is unfolding, with tremendous uncertainty. In particular, international development efforts—accustomed to the donor streams of the approximately past six-to-eight decades from Bretton Woods to USAID—have, for the most part, found the tap not just abruptly and unceremoniously turned down or off, but removed. A different "business model" beyond even "business as usual" is now on order. We contend that three items are now essential to budget into such work moving forward; these points are (1) front-end research to identify in advance (with local communities of practice) what innovations would be relevant, workable, and convenient in a given community before solution design begins, (2) follow-up studies to assess solution *adaptation* (not just adoption) by people in local conditions, and (3) very highly scalable and cost-effective solution delivery through integrated learning-systems approaches.

We are fortunate at Scientific Animations Without Borders (SAWBO) to have thrived over the last decade-and-a-half on a patron–client model more akin to Renaissance artists being commissioned for work by people who wanted media than by an institutionalized, high-level resource dependency on donors. We have also been fortunate to receive grant support for the research needed to build out the ICT learning-systems approach described in this and our earlier book (Bello-Bravo et al., 2023a). We have long argued that research is needed to develop effective ICT-scaling strategies and that scaling efforts provide a basis for learning/research to improve on it—hence, *scaling for research* and *research for scaling* (Bello-Bravo et al., 2023b, p. xiii). Without this bidirectional research, ICT usage risks becoming little better than throwing spaghetti at a wall and hoping some of it sticks.

SAWBO's decade-and-a-half-vetted learning-systems approach lowers the transaction costs of moving gold-standard knowledge across critical links of the value chain (see Chap. 3), i.e., from basic research into the hands of solution developers, from solution developers into the hands of donors wanting to distribute those solutions, and from distributors into the hands of end-recipients who find the solution culturally relevant, comprehensible, and convenient to adopt and use. Its learning-systems model (described in Chap. 2) frontloads the adaptability, flexibility, and resilience for scaling impacts even during this turbulent new era.

As you read this book, keep in mind our argument that a new international development paradigm for achieving the Sustainable Development Goals (SDGs) is emerging: a paradigm in which people's empowerment at the ground level to practice their human right to knowledge (introduced in Chap. 1 and described in Chap. 6) will become an indispensable and necessary cornerstone—a paradigm by which people must be empowered to access knowledge through profoundly cost-effective and massively scaled pathways.

West Lafayette, IN, USA — Julia Bello-Bravo
West Lafayette, IN, USA — Anne Namatsi Lutomia
West Lafayette, IN, USA — John William Medendorp
West Lafayette, IN, USA — Barry Robert Pittendrigh

References

Bello-Bravo, J., Medendorp, J. W., Lutomia, A. N., & Pittendrigh, B. R. (2023a). *Gender, digitalization, and resilience in international development: Failing forward*. Routledge. https://doi.org/10.4324/9781003226383

Bello-Bravo, J., Medendorp, J. W., Lutomia, A. N., & Pittendrigh, B. R. (2023b). Preamble: Why this book now? In J. Bello-Bravo, J. W. Medendorp, A. N. Lutomia, & B. R. Pittendrigh (Eds.), *Gender, digitalization, and resilience in international development: Failing forward* (pp. xi–xiv). Routledge.

DeSilver, D. (2025). *What the data says about U.S. foreign aid*. https://www.pewresearch.org/short-reads/2025/02/06/what-the-data-says-about-us-foreign-aid/

Competing Interests The authors have no competing interests to declare that are relevant to the content of this manuscript.

Contents

List of Figures

List of Table

Chapter 1
Digital Communities of Practice and the Right to Knowledge

The Age of Information

While no exact date would be agreed upon, it is self-evident that a new era—an Information or Digital Age grounded in the emergence of electronic computers from the mid-twentieth century during the Cold War (Bell, 1973)—now shapes a significant part of national economies and people's lives globally. In the United States, the information sector already directly represents approximately $1.5 trillion (6%) of the total gross domestic product (GDP) and was the fourth largest contributor to the 2.5% increase in real GDP in 2023 (BEA, 2024; Bhutada, 2024). These direct contributions do not include the innumerable ways that this information infrastructure supports and informs all other sectors, especially now given the rapidly expanding use of computer-based technologies deemed artificial intelligence (AI), i.e., machine learning, large language modeling processes, neural network simulations, expert systems, and vastly improved robotics. The world is seemingly irreversibly becoming a network of more and less advanced "knowledge societies" (Petrushenko & Vorontsova, 2020; Stehr, 2018; UNESCO, 2005), with all the advantages and disadvantages that accrue to them.

However, access to the predominant benefits and advantages of living in the new knowledge societies accrue most often only to those people who are information literate, live on the "right side" of the digital divide, who speak mainstream global languages, live in urban or peri-urban areas, and are generally not positioned at the bottom of the global economic pyramid. As Nico Stehr (2022) indicates in his most recent work, this new economy is a knowledge economy, driven by "knowledge capitalism," in which gatekeepers of knowledge are in a position not only to profit enormously from the barter of information but also to deliver or deprive others, depending on returns on investment. This profit spawns stratified economies and societies based on one's relative access to the sources of information.

J. Bello-Bravo et al., *The Right to Knowledge*,
https://doi.org/10.1007/978-3-032-03360-4_1

In the following chapters, we emphasize the necessity of making life-improving information accessible to all people at all scales of activity from the national to the personal. This access means being non-exclusive toward people who speak *any* of the world's languages, who may not be print literate in recognized languages, live in rural or isolated areas, or fall on the "wrong sides" of the digital, gender, economic, or educational divides.[1]

While this book addresses gaps that prevent reaching this goal of maximum accessibility—defined in terms of availability, usability, and convenience of information—we appreciate that efforts must be made to reduce or remove those barriers either on the side of the content creators or on the side of the content recipients, or both. It is our view that the provision of access is more effectively focused on the producer side, which provides greater opportunities and resources for bypassing access barriers. For example, one can ask: is accessibility better achieved by expecting end users to be fluent in majority languages to which they are rarely or never exposed? Or by expecting producers to (1) place the information in minority languages spoken by recipients or (2) deliver that information through technological interfaces that accommodate localized contextualization and translation and (3) place them onto platforms and into channels that will reach people where they are?[2]

On this point of translation, automated examples are already cropping up in public interfaces, e.g., as a "translate" option in YouTube comment threads. The quality and accuracy of these automated processes warrant caution and demand for improvements (Hofstadter, 2018). Although all translation is a kind of paraphrase (Bello-Bravo et al., 2023b), unsupervised (automated) translation introduces a heightened risk of less accurate, misleading, or even incorrect information. This problem is especially damaging where scientifically grounded protocols using deliberately precise language are used; loose or inaccurate translations can make the instructions of a protocol ineffective. Moreover, many minority languages are not yet part of the automated-translation catalog, work very poorly, or do not have print transliterations.

Knowledge in the Age of Information

The title of this book addresses a right to *knowledge* and the exercise thereof. Historically, several trajectories around the concept of a right to knowledge exist that intersect or overlap but do not coalesce as yet into any common nomenclature or consensus. Here, we give an orienting introduction to some of these different

[1] Exact tallies of the number of world languages fluctuate from 7000 to approximately 7200, depending on how they are classified and as languages facing extinction die with their last speakers. In this book, we usually refer conservatively to a lower bound of approximately 7000 living languages spoken globally.

[2] Our use of the terms minority and majority languages throughout this book refers only to a numerical majority of speakers (i.e., more widespread or less widespread speakership). No deprecation of "minority" languages is intended.

approaches and our working definition of what a right to knowledge can mean. Chapter 6 describes this right, and its exercise, in greater detail.

Several distinct trajectories can be identified around the idea of the right to knowledge. Most basic is a right to *data*. The right to data is directed primarily at the right to access scientific datasets (Murray-Rust, 2008), especially any generated by public funding. The open data movement was born from this claim, exemplified in the Yale Open Data project. Despite a major focus on data generated by public funds, voices also advocate for access to public and private data collaboratives in the interest of advancing science (Ruijer, 2021). Further, some states require all government-generated data to be available to everyone. However, there are still restrictions in the private sector and academia, where intellectual property rights prevail.

Second, a right to *information* orients toward government transparency and is sometimes also described as the right to know (Janssen, 2012). However, this right to know can include information on environmental hazards, work hazards, medications, and food that is intended to protect against any harmful effects of processes and products. Various "sunshine laws" that require transparency in government are part of this movement, along with consumer protection laws, including informed consent in medicine and product labeling laws.

The more recent epistemic rights movement and its corresponding epistemic justice emphasis also align with this trajectory (Aslama Horowitz et al., 2024). This movement focuses on the attribution of validity to sources of knowledge and any inequalities in epistemic systems that either impose dominant epistemic structures or deprive societies or persons of epistemic alternatives (Byskov, 2021; Cummings et al., 2023; Fricker, 2007). It advocates for "the ability to be treated equally as a knower" (Byskov, 2021, p. 1). Much of the advocacy for the legitimacy of indigenous knowledge falls under this umbrella.

Third, a right to *education* represents the most developed body of thought and is also the most codified in international conventions, like the *Universal Declaration of Human Rights* (Bergström, 2010; Spring, 2000). This line of thought is well-established and may sometimes be referred to as the right to knowledge, but its references are to the right to universal, free, basic education. In some countries, this scope has been extended to secondary education and, in some cases, even to post-secondary education (McCowan, 2011). However, the main focus of this trajectory involves formal education, less so non-formal or informal education. The right to knowledge in this book expressly draws on non-formal and informal education as well as formal education.

Leonardo Burlamaqui (2012) has focused on knowledge *governance*, which moves beyond questions of information access and validation to address the fundamental question of the social obligation to *share* knowledge. This issue concentrates on the balance between intellectual property rights and the general public's right to knowledge. Burlamaqui argues that intellectual property becomes a common good when it reaches a point that all technologies in a given area of expertise depend on it for progress. Once it becomes a public good, it becomes incumbent upon owners possessing intellectual property rights to make such information available to all. Where there is an unwillingness to do so, governance must step in to ensure access.

In this book, a right to knowledge means not only that everyone must obtain access to information and its offshoots, i.e., knowledge and education, but also that those who possess knowledge owe a responsibility to do everything they can to ensure it reaches the people who need it most, especially when it concerns basic human dignity and well-being. This last phrase, as a qualifier on the openness of knowledge generally, is critical. Thus, the knowledge that owners hold an obligation to make freely available is *life-sustaining knowledge* relevant to the "basic" SDGs—knowledge, for example, that supports food and water security (SDG 2 and SDG 6), health and emotional well-being (SDG 3), education (SDG 4), and gender security (SDG 5). Whether *all* knowledge, everywhere and always, could ever be absolutely open in all cases is a matter for the future to decide. Here, the "only" knowledge we insist that people sustain an absolute right to, and for which they must be able to exercise their right, is this life-sustaining knowledge. Chapters 6 and 7 address this premise in more detail.

While meaningful distinctions can be made between data, information, and knowledge that are not merely semantic (Boisot & Canals, 2004; Gray, 2003; Nonaka & Takeuchi, 1997; Wagner & Tacacs, 2021), the pragmatic emphasis of this book focuses more on their governance, accessibility, and convenience for would-be recipients (Bello-Bravo et al., 2022b; Burlamaqui, 2012; Ribot & Peluso, 2003). Convenience is essential. One might recall instances when some knowledge was available, desirable, and usable but still not conveniently enough-to-hand to take up at that particular time, despite a motivated opportunity to use it (Bello-Bravo et al., 2022b).

As with knowledge's availability and usability, we argue that convenience is most strategically addressed on the producer side as well. For one, producers maintain primary control over how information is packaged, disseminated, and contextualized. By prioritizing locally user-accessible interfaces, culturally intuitive designs, and language localization, producers better ensure that information is not simply available but *conveniently* available. Identifying when, where, and through which channels users will be most likely to encounter the information is also data that producers (or their proxies) are best positioned to investigate or produce.

Second, producers can build in reduced time and transaction costs for disseminating knowledge, making access more convenient for end users. Streamlining processes and technologies minimizes the time required for users' information-seeking (Collier & Kimes, 2013). Machine learning algorithms can estimate and preemptively give users better search capacities while reducing the information overload often accompanying digital searches (Carlson, 2003). Of course, these benefits can be diminished by ad-driven algorithms that prioritize paid search results over the ones sought by users (Geradin & Katsifis, 2019).

Producer-side convenience also acknowledges that users vary widely in their degree of access to technology, physical capabilities, and local infrastructures. By recognizing a need in information-system designs for interfaces that are accessible to users with (1) disabilities, (2) low-bandwidth or prohibitively costly Internet connections, or (3) limited technological, educational, or gender-based familiarity with computers, producers can bridge those digital divides and ensure that all people have maximally available access—especially people who have been historically overlooked and marginalized (including women). Making access convenient not only enhances the likelihood of information use but also empowers users with the knowledge to apply that information.

In a previous book (Bello-Bravo et al., 2023a), we addressed the question: *how can information reach billions of people?* Here, we return to some of the technical dimensions of that challenge while concentrating on the broader social problem of making gold-standard, scientifically grounded information accessible to nearly everyone. Fulfilling the right to knowledge involves empowering (digital) communities of practice at every scale—from small groups of change agents to global partnerships—such that they can identify and selectively adapt solutions aligned with global, regional, and local Sustainable Development Goals (SDGs) and benchmarks. Again, the technical solution now in principle exists for achieving such *convenient* usability of knowledge by billions of people social challenges around doing so remain.

Pragmatically emphasizing producer-side convenience moves toward solving this social challenge, since its locus involves not only *how* to make information available, usable, and convenient for essentially anyone but also *why*. Given that "the core element that drives the digital society is digitalized information" (Park, 2017, p. 161), powerful actors within existing social hierarchies can be averse to yielding ground or sharing power; this reluctance is a core aspect of the social problem yet to be solved, which ultimately involves both producers and recipients recognizing the rights *and* responsibilities around any exercises of the right to knowledge (Kapczynski, 2008). A primary goal of this book is to enable the exercise of that right for all people at all times.

Digital Communities of Practice and the SDGs

In this time of dramatic reductions to resources for international development, to continue to make strides toward the SDGs requires moving beyond the "giving man a fish" and "teaching a man to fish" models. Instead, efforts must be geared toward empowering people to "teach themselves to fish." This change of emphasis is not a call for a moratorium on the older approaches. If organizations can still provide direct material assistance ("give fish") or training to meet the needs of the world's poor ("teach to fish"), these efforts are still urgently needed and welcomed, so long as their innovations provide maximally effective short-term interventions with the least-detrimental long-term effects. In this new era, we may hear voices who sharply criticize and call for an end to such past efforts, without offering an alternative. This amounts to unethically abandoning the world's poor. Equally, to stand back and say, "Let them teach themselves" does not suffice. We still have the mandate and responsibility to the world's poor to stage the conditions by which they can "teach themselves to fish."

On "our" side—on the humanitarian side of the commitment of international development to meet the needs of the world's poor—one way to do so involves mass-scaling a vast, scientifically grounded encyclopedia of digital knowledge reachable in every corner of the world, an "encyclopedia" that people can access, use, and find relevant solutions to problems they face, above all around life-sustaining knowledge. SAWBO's body of work is already dedicated to this end (Bello-Bravo et al., 2023a). On "their" side, this objective requires change agents and organizations who can (1) utilize such encyclopedic knowledge via demand-driven, low-cost learning systems, and (2) empower access by entire cross-sections of the population to large-scale digital knowledge in as many diverse channels as possible. Such an arrangement can circumvent traditional gatekeepers of knowledge (thus, again, partially solving the social problem of knowledge access), so long as these systems obtain a measure of self-sustainability and operate independently of grants or external funding.

A central touchstone in this book, then, are these *digital communities of practice*, which can empower people to exercise their human right to knowledge and obtain solutions to problems they have recognized in their communities, nations, regions, or even globally. While these digital communities of practice are not strictly or only grassroots organizations, we draw on how communities of practice can resemble grassroots organizations, which are already recognized as essential for attaining the SDGs (Flores & Samuel, 2019; Islam, 2021). Per Flores and Samuel (2019), grassroots organizations often attract people who are otherwise disenfranchised or left out of development programs, thus leading to stronger communities. Prior to international development's current seismic shocks, grassroots organizations were foundational, a literal base, especially when linked to funding for specific projects during specific time intervals. Ideally, this foundation includes the nodes and connections within fully capacitated networks spanning the three societal sectors (state, civil society, and capital) (Roberts et al., 2005).

SAWBO Partnership Frameworks (SDG 17)

Partnerships are critical for meeting the SDGs. SAWBO team members have considerable experience in the conventional structures, but their partnership practices more broadly draw on four strategic lenses: learning systems, virtual and hyper-collaboration, network theory, and the theory of change model (USAID, 2016). In general, the theory of change framework offers a road map for how partnerships achieve impacts—in SAWBO's case, by distributing free, scientifically validated animated videos on life-sustaining knowledge linguistically tailored to reach diverse socio-demographically audiences. Additionally, the learning-systems approach to partnership supports the development of flexible, scalable mechanisms for sharing knowledge effectively across varied contexts.

SAWBO's animated videos offer a low-cost, highly responsive, and flexible method for adapting educational content to fit varying linguistic and cultural settings, making them especially effective for reaching underserved and rural communities, enhancing local agency, and aligning with SDGs. SAWBO's emphasis on virtual and hyper-collaborative methods across international initiatives facilitates fast, targeted communication, enabling timely production of content during crises like environmental disasters or pandemics. During COVID-19 pandemic, for instance, mobility restrictions led to increased reliance on digital tools like WhatsApp to maintain direct engagement with farmers and support real-time knowledge exchange and troubleshooting. SAWBO's partnerships also function through network-based structures with three defined tiers: active, task/project completed, and enabled. These tiers support long-term cooperation, ongoing distribution of educational videos, and the cultivation of resilient, collaboration-driven systems for the future.

Active Partnerships: Active partnerships refer to collaborative engagements in which all participating entities are consistently involved in joint activities, open communication, and shared initiatives. These relationships are characterized by continuous interaction, mutual contribution of resources, and the ongoing pursuit of common objectives. Within a network framework, nodes in the active partnership zone represent the dynamic actors—such as individuals, institutions, or teams—currently contributing to and shaping the project. These nodes embody the strong, functioning connections that define an active partnership.

Task or Project-Completed Partnerships; This partnership type describes collaborations that were previously active but are now dormant due to the completion of the shared initiative. Although the relationship still exists, ongoing communication and engagement are minimal or paused. These partnerships remain viable and can be reactivated if future needs arise. In the network model, nodes within this category represent recent collaborators—partners who previously worked together and may still utilize the video animations, though usage might have declined as their specific project has concluded. Examples include partnerships formed during response efforts for Ebola and COVID-19.

Enabled Partnerships: These interactions encompass groups or people who have received SAWBO's animations and may or may not be using them currently. They include both formal and informal collaborators, including organizations that have independently adopted SAWBO's content without prior direct engagement. In some cases, SAWBO only becomes aware of these partnerships after an organization reaches out unexpectedly to share that they have been using the videos. Nodes in the enabled partnership zone represent entities that are part of the broader network but not actively engaged—often functioning on the periphery or as secondary participants. These connections are not currently active but remain potential sites for future collaboration. Often, these partners take the animations and distribute them widely, sometimes without informing SAWBO, enabling extended reach through downstream sharing. These partnerships reflect a model of future collaboration: resilient, self-sustaining, and not dependent on constant oversight, where all tiers can be accessed by digital communities of practice to begin, continue, or extend knowledge access (Lutomia, 2024).

Tools for Capacity Building

Utilizing video animations as a tool for capacity building offers a dynamic and effective method for SAWBO's partnerships to empower people, facilitate knowledge sharing, and foster sustainable practices. This approach has been instrumental in advancing several key SDGs, including No Poverty (SDG 1), Zero Hunger (SDG 2), Good Health and Well-Being (SDG 3), Quality Education (SDG 4), Gender Equality (SDG 5), Reduced Inequalities (SDG 10), and Partnerships for the Goals (SDG 17). Table 1.1 summarizes the average number of SDG indicators and percentage of video content related to all 17 SDGs in an excerpted selection of 71 SAWBO videos.[3]

Table 1.1 Average number of indicators and percentage of video content in select SAWBO videos ($n = 71$)

SDG	Goal	Average # of indicators per video	% of video content addressing goals
SDG 1	No Poverty	2.17	30.99
SDG 2	Zero Hunger	2.62	32.75
SDG 3	Good Health and Well-being	3.28	27.35
SDG 4	Quality Education	3.68	36.76
SDG 5	Gender Equality	1.48	16.43
SDG 6	Clean Water and Sanitation	1.32	16.55
SDG 7	Affordable and Clean Energy	0.32	5.40
SDG 8	Decent Work and Economic Growth	2.52	21.01
SDG 9	Industry, Innovation and Infrastructure	3.48	43.49
SDG 10	Reduced Inequalities	1.31	13.10
SDG 11	Sustainable Cities and Communities	1.79	17.89
SDG 12	Responsible Consumption and Production	2.46	22.41
SDG 13	Climate Action	1.62	32.39
SDG 14	Life Below Water	0.0	0.0
SDG 15	Life on Land	0.87	7.28
SDG 16	Peace, Justice and Strong Institutions	2.20	18.31
SDG 17	Partnerships for the Goals	3.69	19.42

Reprinted and adapted with permission from Rodríguez-Domenech et al. (2023)

[3] There is no data for SDG14 (Life Below Water) because, to date, SAWBO has not been commissioned to produce content for that SDG (Rodríguez-Domenech et al., 2023).

Communities of Practice

The notion of *digital* communities of practice is anchored in Lave and Wenger's (1991) original formulation of *communities of practice*, described as a group of people—often from dissimilar backgrounds, united principally by their shared interest in a topic—who collectively study, experiment, and seek to master a knowledge and practice of a particular phenomenon. Often, communities of practice facilitate collectively sharing best practices and peer learning. By engaging in communities of practice, smallholder farmers can draw on contemporary and traditional knowledge to enhance their skills, access new technologies, and build social capital, all of which contribute to the attainment of the SDGs (Wenger, 1998a).

Lave and Wenger (1991) expressly framed this process as *situated learning*, contextualized specifically by participatory structures and processes with like-minded others. Whether through formal, informal, or non-formal means, it is important that the self- and collectively directed learning of a community of practice is intrinsically motivated; that is, it is driven by a "disinterested" and empirical inquiry into a contextually grounded topic, with the aim of enabling members to become local knowledge experts (Bradbury & Middlemiss, 2015).

It is especially in this sense that being able to exercise the right to knowledge becomes essential, since changes to knowledge—and thus practice—are the core activity of a community of practice. This change necessarily means that knowledge is never static or settled but dynamic and adaptive as new insights are gained. While this adaptation requires processes for vetting, adapting, and curating each iteration of knowledge, along with its practices and artifacts, we should remain agnostic about the pathways through which content is made available. Despite the current condition of international development, the world remains on a trajectory of increasing access to tools that allow people to push, pull, and use such content (Ngonga et al., 2024). There is no "universal" or "single" method for use and scaling across countries, communities, and time. Rather, the constant is that the recognition of everyone's right to access life-sustaining knowledge is a global public good. Realizing that right means prioritizing content that supports the SDGs and making sure that it is delivered in formats that people can understand—languages they speak comfortably and media they can access, whether or not they are print-literate—especially content that addresses basic human needs, including dignity (de Oliveira Cintra Farias & Ferreira, 2024).

Further, as a form of institution or organization, a community of practice can become susceptible to problems characteristic of institutions, including internal power struggles, inertia, ineffective leadership, emergent disagreements about "schools of thought," and challenges around sustaining the community when enthusiasm wanes, disagreements overflow, attrition occurs, or bureaucratic succession becomes necessary, especially in the integration and orientation of newcomers (Stuckey & Smith, 2004). Because communities of practice often originate informally, they may resist or be unfamiliar with the structures required to formally organize their work or to navigate the downstream hurdles of engaging other organizations, collaboratives, or partnerships (be those other communities of practice or not) (Wenger, 1998b). Even so, as associations with the core skill of focusing on a topic to master it, communities of practice may be uniquely disposed to exploring and resolving these institutional challenges.

Communities of Practice in Smallholder Agriculture

The right to knowledge in smallholder agriculture involves ensuring farmers' access to essential information and resources to make informed decisions. This right includes access to agricultural research, market information, weather data, and best practices for sustainable farming. For smallholder farmers, especially those in remote or under-resourced areas, traditional means for accessing such knowledge can be limited by little or no print-literacy, lack of digital or electrical infrastructures, and digital divides (Mudege et al., 2015). Nevertheless, learning-systems approaches using video animations have shown a capacity to overcome these barriers (Bello-Bravo et al., 2023a).

Three aspects of participation emphasized by Kuh (2012) include engagement, imagination, and alignment with goals. Digitalization of communities of practice radically reorients their participatory capacity toward providing solutions to communities locally and supporting the SDGs globally. As such, digital communities of practice form the through-line in the following chapters of this book, centered expressly on empowering *all* people to be able to exercise their human right to knowledge and thus contribute to the benchmarks for meeting the SDGs.

While Chap. 2 describes SAWBO's flexible, adaptable, and resilient learning-systems approach for moving gold-standard knowledge "off the shelf" through the production-side links of the value chain into the hands of recipients who want it and can most benefit from it, Chap. 3 then demonstrates how loosely coupling the transaction costs involved in bridging those links can *decrease* those costs with greater scaling (Bello-Bravo et al., 2022a). This reduction not only enables end-recipients to select or adopt gold-standard innovations that address problems in their communities but also allows solution designers to decide which basic research to move off the shelf into development; it also helps agenda-setters determine which solutions to deploy in the field. Chapter 4 then discusses the radical potential for mass-scaling ICTs, making it no exaggeration to suggest that the technical problem of delivering knowledge to billions has largely been solved. Chapter 5 anticipates the expanding possibilities of big data, machine learning, and AI for supporting ICT mass-scaling in the service of the SDGs.

Yet even as the technical problem of ICT mass-scaling is solved in principle, "wicked" social problems constraining its real-world implementation remain. Chapters 6 and 7 address these by further advocating the right to knowledge in its legal and everyday socio-pragmatic sense—including how that right is empowered or obstructed by the language in which knowledge is encoded—and the role of sustainability. These issues have a deep historical resonance, especially given marginalization and erasure of indigenous knowledge and practices. Since indigenous communities often resemble communities of practice themselves, any exclusion of their sustainable knowledge and practices represents a profound loss for humanity as a whole and for achieving the SDGs specifically.

One of the central emphases in the *Brundtland Report*'s (1987) definition of sustainability is the imperative to give overriding priority to meeting the needs of the world's poorest. In light of current shocks to the international development sector, it is especially urgent to empower the world's poorest to find, select, and implement innovations that meet any needs that, until recently, were addressed (if at all) by large-scale humanitarian aid coalitions.

Digitally empowering communities of practice is one highly cost-effective way, even under the current constraints, to empower people as needed. The critical difference and offer of this approach is the *bidirectional* interface between a loosely coupled, locally situated, learning-interested community of practice and a globally positioned, learning-systems approach to knowledge. This arrangement means that "we" are not simply all-knowing producers or teachers giving recipients fish or teaching students to fish; we are students as well, and they are teachers. Not only are they empowered to teach themselves to fish, but they can also teach us what solutions they need or have discovered. They can become participants in setting the agenda.

References

Aslama Horowitz, M., Nieminen, H., Lehtisaari, K., & D'Arma, A. (Eds.). (2024). *Epistemic rights in the era of digital disruption*. Springer Nature.

BEA. (2024). *Contribution to percent change in real GDP by industry group, 2023*. https://www.bea.gov/system/files/gdp4q23-3rd-chart-05.png

Bell, D. (1973). *The coming of post-industrial society: A venture in social forecasting*. Basic Books.

Bello-Bravo, J., Medendorp, J. W., Lutomia, A. N., Reeves, N. P., Tamò, M., & Pittendrigh, B. R. (2022a). Dramatically increased accessibility and decreased cost-per-person impacts are needed for scaling IPM in Africa. *Current Opinion in Insect Science, 54*, 100971. https://doi.org/10.1016/j.cois.2022.100971

Bello-Bravo, J., Muyodi, F. J., Nalwanga, R., Nakafeero, C. M., & Rutechura, F. R. (2022b). Time is not always money: A preliminary study on socially sustainable strategies banana xanthomonas wilts (BXW) mitigation efforts in Mbarara region Uganda. *Journal of Agricultural Extension and Rural Development, 14*(2), 90–101. https://doi.org/10.5897/JAERD2022.1319

Bello-Bravo, J., Medendorp, J. W., Lutomia, A. N., & Pittendrigh, B. R. (2023a). *Gender, digitalization, and resilience in international development: Failing forward*. Routledge. https://doi.org/10.4324/9781003226383

Bello-Bravo, J., Medendorp, J. W., Lutomia, A. N., & Pittendrigh, B. R. (2023b). Translating to connect local languages for learning. In *Gender, digitalization, and resilience in international development: Failing forward* (pp. 234–254). Routledge. https://doi.org/10.4324/9781003226383

Bergström, Y. (2010). The universal right to education: Freedom, equality and fraternity. *Studies in Philosophy and Education, 29*, 167–182. https://doi.org/10.1007/s11217-009-9174-y

Bhutada, G. (2024). *Visualizing U.S. GDP by industry in 2023*. https://www.visualcapitalist.com/visualizing-u-s-gdp-by-industry-in-2023/

Boisot, M., & Canals, A. (2004). Data, information and knowledge: Have we got it right? *Journal of Evolutionary Economics, 14*, 43–67. https://doi.org/10.1007/s00191-003-0181-9

Bradbury, S., & Middlemiss, L. (2015). The role of learning in sustainable communities of practice. *Local Environment, 20*(7), 796–810. https://doi.org/10.1080/13549839.2013.872091

Burlamaqui, L. (2012). Knowledge governance: An analytical approach and its policy implications. In L. Burlamaqui, A. C. Castro, & R. Kattel (Eds.), *Knowledge governance: Reasserting the public interest* (pp. 3–26). Anthem.

Byskov, M. F. (2021). What makes epistemic injustice an "injustice"? *Journal of Social Philosophy, 52*(1), 114–131. https://doi.org/10.1111/josp.12348

Carlson, C. N. (2003). Information overload, retrieval strategies and internet user empowerment. In L. Haddon (Ed.), *The good, the bad and the irrelevant (COST 269)* (pp. 169–173). Media Lab UIAH.

Collier, J. E., & Kimes, S. E. (2013). Only if it is convenient: Understanding how convenience influences self-service technology evaluation. *Journal of Service Research, 16*(1), 39–51. https://doi.org/10.1177/1094670512458454

Cummings, S., Dhewa, C., Kemboi, G., & Young, S. (2023). Doing epistemic justice in sustainable development: Applying the philosophical concept of epistemic injustice to the real world. *Sustainable Development, 31*(3), 1965–1977. https://doi.org/10.1002/sd.2497

de Oliveira Cintra Farias, F., & Ferreira, H. L. P. (2024). How much for human dignity? A Kant-Maslow compared approach to the basic conditions of life. *Revista Juridica da UFERSA, 8*(15), 136–154.

Flores, W., & Samuel, J. (2019). Grassroots organisations and the sustainable development goals: No one left behind? *BMJ, 365*, l2269. https://doi.org/10.1136/bmj.l2269

Fricker, M. (2007). *Epistemic justice: Power and the politics of knowledge*. Oxford University Press.

Geradin, D., & Katsifis, D. (2019). An EU competition law analysis of online display advertising in the programmatic age. *European Competition Journal, 15*(1), 55–96. https://doi.org/10.1080/17441056.2019.1574440

Gray, R. (2003). A brief historical review of the development of the distinction between data and information in the information systems literature. In *Proceedings of the Americas Conference on Information Systems (AMCIS 2003)* (pp. 2856–2862). Association for Information Systems.

Hofstadter, D. (2018). *The shallowness of Google Translate*. https://www.theatlantic.com/technology/archive/2018/01/the-shallowness-of-google-translate/551570/

Islam, M. T. (2021). Grassroots political movement to achieve SDGs in developing countries. In W. Leal Filho, A. Marisa Azul, L. Brandli, A. Lange Salvia, P. G. Özuyar, & T. Wall (Eds.), *Peace, justice and strong institutions* (pp. 348–356). Springer.

Janssen, K. (2012). Open government data and the right to information: Opportunities and obstacles. *The Journal of Community Informatics, 8*(2). https://doi.org/10.15353/joci.v8i2.3042

Kapczynski, A. (2008). *The access to knowledge mobilization and the new politics of intellectual property*. https://www.yalelawjournal.org/forum/access-to-knowledge-mobilization-pocket-part

Kuh, L. P. (2012). Promoting communities of practice and parallel process in early childhood settings. *Journal of Early Childhood Teacher Education, 33*(1), 19–37. https://doi.org/10.1080/10901027.2011.650787

Lave, J., & Wenger, E. (1991). *Situated learning: Legitimate peripheral participation*. Cambridge University Press.

Lutomia, A. N. (2024). Co-designing future-based partnership for a university-affiliated organization using a futuristic approach. *Interdisciplinary Journal of Partnership Studies, 11*(2), 7. https://doi.org/10.24926/ijps.v11i2.6421

McCowan, T. (2011). Human rights, capabilities and the normative basis of 'education for all'. *Theory and Research in Education, 9*(3), 283–298. https://doi.org/10.1177/1477878511419566

Mudege, N. N., Nyekanyeka, T., Kapalasa, E., Chevo, T., & Demo, P. (2015). Understanding collective action and women's empowerment in potato farmer groups in Ntcheu and Dedza in Malawi. *Journal of Rural Studies, 42*, 91–101. https://doi.org/10.1016/j.jrurstud.2015.09.002

Murray-Rust, P. (2008). Open data in science. *Nature Precedings*. https://doi.org/10.1038/npre.2008.1526.1

Ngonga, E., Ombok, B., Osewe, D., Aila, F., & Odhiambo, G. (2024). The role of multi actor communities of practice in dissemination of push-pull technology. *Journal of Agriculture, Science and Technology, 23*(3), 166–174. https://doi.org/10.4314/jagst.v23i3.10

Nonaka, I., & Takeuchi, H. (1997). *The knowledge-creating company: How Japanese companies create the dynamics of innovation*. Oxford University Press.

Park, S. (2017). Information is power. In *Digital capital* (pp. 161–183). Palgrave Macmillan.

Petrushenko, Y., & Vorontsova, A. (2020). Knowledge society: The evolution of the concept in the context of achieving sustainable development goals. In W. Leal Filho, A. M. Azul, L. Brandli, P. G. Özuyar, & T. Wall (Eds.), *Quality education* (pp. 511–525). Springer. https://doi.org/10.1007/978-3-319-95870-5_40

Ribot, J., & Peluso, N. L. (2003). A theory of access. *Rural Sociology, 68*(2), 153–181. https://doi.org/10.1111/j.1549-0831.2003.tb00133.x

Roberts, S. M., Jones, J. P., III, & Fröhling, O. (2005). NGOs and the globalization of managerialism: A research framework. *World Development, 33*(11), 1845–1864. https://doi.org/10.1016/j.worlddev.2005.07.004

Rodríguez-Domenech, M. Á., Bello-Bravo, J., Lutomia, A. N., Medendorp, J. W., & Pittendrigh, B. R. (2023). Digital education and community outreach for sustainable development in Africa. In W. Leal Filho, I. R. Abubakar, I. da Silva, R. Pretorius, & K. Tarabieh (Eds.), *SDGs in Africa and the Middle East Region: Implementing the UN sustainable development goals—Regional perspectives* (pp. 1–31). Springer. https://doi.org/10.1007/978-3-030-91260-4_65-1

Ruijer, E. (2021). Designing and implementing data collaboratives: A governance perspective. *Government Information Quarterly, 38*(4), 101612. https://doi.org/10.1016/j.giq.2021.101612

Spring, A. (Ed.). (2000). *Women farmers and commercial ventures: Increasing food security in developing countries*. Lynne Rienner Publishers.

Stehr, N. (2018). Modern societies as knowledge societies. In M. T. Adolf (Ed.), *Nico Stehr: Pioneer in the theory of society and knowledge* (pp. 309–331). Springer International Publishing. https://doi.org/10.1007/978-3-319-76995-0_20

Stehr, N. (2022). *Knowledge capitalism*. Routledge.

Stuckey, B., & Smith, J. D. (2004). Building sustainable communities of practice. In P. Hildreth & C. Kimble (Eds.), *Knowledge networks: Innovation through communities of practice* (pp. 150–164). IGI Global.

UNESCO. (2005). *Towards Knowledge Societies (UNESCO World Report Document SHS.2005/WS/44)*. UNESCO.

USAID. (2016). *An analysis of theories of change in USAID solicitations for education programs in crisis and conflict affected environments*. USAID.

Wagner, I., & Tacacs, Z. (2021). Data-as-a-service versus information-as-a-service: Critical differences in theory, implementation, and applicability of two growing cloud services. In N. Kryvinska & A. Poniszewska-Marańda (Eds.), *Developments in information & knowledge management for business applications* (Vol. 2, pp. 113–149). Springer.

Wenger, E. (1998a). Communities of practice: Learning as a social system. *Systems Thinker, 9*(5), 1–10. https://doi.org/10.1017/CBO9780511803932

Wenger, E. (1998b). *Communities of practice: Learning, meaning, and identity*. Cambridge University Press.

Chapter 2
Learning-Systems Approaches for the Right to Knowledge: Highly Accessible Materials for Achieving the SDGs

Introduction

Ongoing impacts from COVID-19, various losses of global food security, setbacks in progress or limited gains toward the SDGs (particularly regarding gender), and the current seismic shocks reverberating in international development, all raise how and if the SDGs can be realized. These reversals and shortfalls highlight the ever-urgent need for innovations in strategy, research, and planning—toward either reaching the SDGs, resetting or revising them, or starting anew from scratch. In particular, setbacks to basic life necessities—above all, food, water, shelter, and bodily autonomy, which are put at greater risk by climate impacts—underscore an obvious vulnerability of local and global systems to disruption; moreover, the limited progress toward achieving gender equality for wholly half of the human population is less recognized in its urgency, losses, and squandering of opportunity to increase women's participation, yet still urgent in terms of human well-being. Here, too, progress has been slow and easily reversed.

While positive outcomes were definitively accomplished ahead of schedule for reducing extreme poverty by half (SDG1), the subsequent precarity and reversal of those achievements during COVID-19 are not merely historically unfortunate; they are life threatening for millions on the margins (Bhattacharya & Bose, 2023). Despite progress, much of that has proven insufficiently resilient facing totality of complex and interconnected global challenges, with many of the usual groups still being overlooked (Haji & Himpel, 2024). This inconsistency again points to the need for a different systematic approach for tackling root causes of basic necessity shortfalls and discovering more sustainable solutions that can be scaled up globally. Among those solutions must be included an expansion of the knowledge base of the world's poorest communities. Not only does basic, life-sustaining knowledge in agriculture, nutrition, and health give people the means to cope during crises, but it

J. Bello-Bravo et al., *The Right to Knowledge*,
https://doi.org/10.1007/978-3-032-03360-4_2

also allows them to innovate based on reliable, scientifically based solutions that address root causes from the ground up, even if external support mechanisms fail.

By way of analogy, the model of a fire department—or emergency response systems more broadly—is most effective when designed to prevent emergencies, not merely respond to them. A fire department should be the last, not the first, line of defense. Safer building materials, effective fire codes, home alarms, early alert systems, and above all, widespread public awareness of best practices form the real foundation of fire safety. Fire departments stand by precisely because these measures may fall short.

Public knowledge of these practices—not just by residents at home but also public officials, system designers, and researchers—is essential. Scaling this awareness is, in principle, a solved technical problem; what remains are the political and ethical demand to secure the right to knowledge so that actionable information reaches everyone. Public education programs—delivered through schools, announcements, and community events—build the kind of resilience that saves lives. Withholding such life-saving knowledge, when its benefits are clear and its dissemination feasible, is indefensible.

The same holds true for knowledge related to life's basic necessities: water, shelter, food, health, bodily autonomy, and dignity. These fundamentals are not just needs in times of crisis but foundations of resilience itself. The SDGs extend and specify the *Universal Declaration of Human Rights*, adopted in 1948 (Winkler & Williams, 2017), such that to deny entire populations access to essential, life-sustaining knowledge is not only unconscionable but arguably a violation of the Declaration's spirit. From this human rights perspective, we must prioritize the needs of the world's most vulnerable (Brundtland et al., 1987, p. 41), beginning with the right to access and act on knowledge vital to food, water, shelter, health, and personal safety. By securing sovereignty over these essentials—aligned with the SDG benchmarks—we can realize the broader social equity goals embedded in sustainable development and global justice movements.

Until recently, one could argue that we did not have the means to reach every human being on the planet, but with the advance of information and communication technologies (ICTs), that window of unreachability is rapidly vanishing (Shanahan, 2024). Creating platforms that take advantage of recent advances in ICT to dramatically expand the dissemination of essential life-sustaining information is now within our reach. The question only remains, how do we grasp it.

Must we accept that these issues will persist indefinitely, or can we, as international development practitioners, work toward making our roles obsolete? Treating these problems as permanently unsolvable is ultimately not sustainable. While some are "wicked" challenges that resist definitive solutions, the SDGs were created to set concrete, attainable targets for human well-being. Achieving those targets is impossible without a foundational knowledge base available to all.

Our goal can be to create conditions in which our interventions are no longer needed—or at least shift toward enabling new emergences rather than constantly remediating recurring emergencies. Crises will still occur, as no resilience expert can anticipate every scenario. But effective responses will always depend on the capacity to generate creative solutions to unanticipated problems. That capacity, in turn, requires broad access to validated knowledge.

We can ensure that even the most remote communities have access to an "encyclopedia" of scientifically validated research-for-development knowledge—in their own languages, accessible without the need for print-literacy, and available in real-time. This access will allow people to respond rapidly and effectively to the challenges that they face. Such knowledge systems must serve not only majority-language speakers but also the approximately 7000 local-language communities worldwide.

Achieving this objective will require a different, systematic approach—not only for creating such content but also for sharing it in formats that are easy to access and use by end users and local innovators. This process is how we fulfill the right to knowledge: by ensuring that life-sustaining, problem-solving information is within reach of all, regardless of language, literacy, or geography.

Given the systemic barriers preventing on-the-ground realization of the SDGs for certain communities, it no longer suffices only to "work within the system" to overcome them, as the system itself can become the barrier. This obstruction includes discriminatory policies, unequal resource distribution, and culturally restricted "solutions" for ameliorating these problems. These flawed systems not only hinder the realization of the SDGs but also exacerbate the problem and lay the ground for their recurrence. For example, the democratic concept of majority rule will inevitably produce minority exclusion unless specific measures are taken to reverse those trends. Although not an intentional part of democratic societies, constant effort is required to avoid lapsing into injustice for excluded marginalized communities (Agarin, 2020; Lublin & Bowler, 2018; Rawls, 1971).

Recognizing the right to life-sustaining knowledge thus becomes a fundamental component for addressing global and local challenges for realizing the SDGs (Vargas & Lee, 2018). Some have already advocated for an 18th SDG, articulating a right to information and communication (Vargas & Lee, 2023). Without this basic right, the future that we want becomes difficult to realize. Unless all groups can attain access to information and actionable knowledge, equality remains out of reach. The benefits of the right to essential knowledge must be shared equally, enabling individuals and communities to make informed decisions, adopt sustainable practices, and advocate for their own rights. In the context of climate change, mass dissemination of knowledge about sustainable technologies and ecological stewardship is critical to driving collective action and securing *a world with a future*—one that fulfills the core principle of sustainability: meeting present needs without compromising future generations' ability to do so as well (Brundtland et al., 1987).

Ensuring the right to life-sustaining knowledge involves opening new channels of information able to circumvent intentional and unintentional systemic barriers restricting access to information knowledge, whether digital divides, educational disparities, censorship, or research methodologies that reproduce, rather than resist, current barriers to access. By prioritizing the right to knowledge, we facilitate less exclusive and more effective strategies—ultimately not relying on reactive interventions and instead fostering more proactive and equitably sustainable interventions for the future.

Learning-Systems Approaches Are Needed to Reach the SDGs

That a systems approach is necessary for achieving the SDGs is well-established; less well-established is what such a systems approach must look like. This book builds on prior findings and theorizing that centers on a *learning-systems approach* (Bello-Bravo et al., 2023).

Long ago, Hollingworth (1932) identified *change of behavior* as the fundamental evidence of learning; he added to this basic insight that the *speed* with which that change of behavior is realized is critical. This speed is doubly important for governmental, scientific, and administrative efforts to affect changes of behavior that support the SDGs, because many of societies' existing structures to deal with crises are notoriously slow (Guha & Chakrabarti, 2019; Meuleman, 2022). Further, the complexity of interlocking global systems requires a comprehensive framework to account for the interrelated activity of various sectors and issues. The SDGs offer one such framework, particularly through SDG17's emphasis on partnerships. This systems approach helps ensure that progress in one area supports, rather than undermines, progress in others. By integrating learning—understood as behavioral change occurring on faster time-scales—the SDGs become more adaptive, responsive, and ultimately effective.

For example, during the COVID-19 pandemic, 5592 survey respondents in select regions of Kenya, Nigeria, and Ghana were asked about a mass-disseminated video depicting a jerrycan technique for preventing postharvest loss and improving food security. While 56% reported never having previously seen a video on the technique, 93.2% of those people who did see the video indicated they now understood how to use it, 90.3% stated they could teach it to others, and 58.1% self-reported using it after watching the video. The reach, engagement, and self-reported use of his life-sustaining knowledge for improved food security was not accomplished with a complete plan in advance but involved collaborative flexibility from the outset. Some of the video distribution channels envisioned for use at the outset proved unusable; alternatives were successfully sought and utilized. Overall, this breakthrough presented an unusual opportunity brought about by the extraordinary situation of COVID-19 but clearly illustrates what a learning-systems approach can look like for achieving very high levels of reach, engagement, and self-reported use through mass-scaling of life-sustaining knowledge. It is precisely such innovative and alternative approaches to address the right to knowledge that are needed if the SDGs "future we want" is to become a reality.

National governments, development organizations, universities, and other entities recognize the SDGs as critical for promoting and achieving sustainable development across all sectors (Azcona et al., 2020; Bello-Bravo & Lutomia, 2020; Gates, 2020; Mukarram, 2020). This broad endorsement of the SDGs by diverse stakeholders highlights their importance and relevance and makes them an obvious channel for enhancing them via a learning-systems approach. Governmental integration of the SDGs into national policies and development plans and developmental organizations' alignment of their programs with these global targets affords the needed scale of reach for learning approaches to be effective. Universities especially contribute through research, innovation, and capacity building, not only enabling knowledge and expertise that support sustainable development efforts but exploration and development of methodologies that draw on learning systems.

The SAWBO Learning-Systems Approach

Prior work established how SAWBO's content aligns with the SDGs (Rodríguez-Domenech et al., 2019; Rodríguez-Domenech et al., 2023), demonstrating practical applications in various SDG contexts, showcasing the reach, engagement, and impact of educational interventions on community development. By aligning content with the SDGs, SAWBO's learning-systems approach lays a base of knowledge for global sustainability goals. These objectives highlight effective learning-systems strategies for advancing the SDGs and the broader right to mass-scaled access to life-sustaining knowledge, particularly within the context of a global public good e-extension system. SAWBO serves as both an academic and pragmatic model for how encyclopedias of scientifically validated research-for-development knowledge can be created, translated into as many languages as possible—ideally all of them—and shared widely with economically marginalized and remote populations.

To date, SAWBO's video library has produced content in over 300 languages and dialects, covering all but one of the 17 SDGs (SDG 14: Life Below Water). This reach demonstrates the versatility of SAWBO's approach and its trajectory toward building a truly inclusive, accessible encyclopedia of research-for-development (R4D) innovations. The content draws on research from a wide range of institutions—including National Agricultural Research Organizations (NAROs), Consultative Group for International Agricultural Research (CGIAR) centers, universities, and more—and aims to make scientifically validated solutions accessible globally.

Although SAWBO's content addresses core areas, including food security, health, education, and gender security, the content arises in response to needs articulated by funders and collaborators; in this way, SAWBO is solution-oriented but not product-centric. Its growing library supports a "menu of solutions" that can be adopted or adapted by actors at any scale—from individual entrepreneurs to governments and global coalitions—depending on local relevance and priorities. Each new funding mechanism is an opportunity to expand this encyclopedia by adding topic areas, new languages, or both. The guiding principle behind this model is accessibility: encyclopedias exist so that users can access specific knowledge at the time and place they need it—just as one would look up "cats" in a volume labeled "C." This logic supports "pull" strategies in international development, where user-driven demand determines what information is produced and accessed. However, in many marginalized contexts, the lack of access infrastructure renders such pull strategies infeasible. That shortage must change. New platforms must be and are being built, with content deployed to fulfill the right to knowledge.

SAWBO's most frequently addressed SDGs include SDG 9 (Industry, Innovation, and Infrastructure), SDG 4 (Quality Education), SDG 1 (No Poverty), SDG 2 (Zero Hunger), SDG 3 (Good Health and Well-being), and SDG 5 (Gender Equality). This book's emphasis on SDGs 1–5 reflects a focus on life-essential knowledge; SDG 4 highlights the foundational role of education, while SDG 9 emphasizes the infrastructure needed to make learning and access to knowledge possible on a global scale.

The ultimate objective is to build an encyclopedia of R4D innovations that is easily accessible, highly cost-effective, and usable by anyone, regardless of language, literacy level, socioeconomic status, or rural/urban geographic location. To achieve this, learning systems must not only generate high-quality content but also create effective mechanisms for distributing it and ensuring that both end users and knowledge producers can learn from and contribute to the system.

Learning-Systems Approaches

Learning-systems approaches differ from other systems approaches in three key ways: (1) by being proactively adaptive, not just reactive, (2) by treating knowledge as accessible, localized, and dynamically iterative, (3) and their suitability to support people's exercise of the right to knowledge, especially when underserved or with limited resources. *Learning* in this context means adopting a different attitude toward what technological solutions look like (Bello-Bravo, 2023; Daré et al., 2014).

A central way to define learning-systems approaches is not by their complexity (which they may be) but their capacity to transform. Unlike conventional systems frameworks that incorporate feedback to correct behavior, learning systems integrate feedback in ways that *transform* the system's structure, trajectory, and even purpose (Davies & Nutley, 2018). This ability of a system to learn is not a peripheral enhancement; it is the central function that enables responsiveness, resilience, and ultimately, relevance in the face of complex global challenges like the SDGs.

Rooted in diverse frameworks across engineering, organizational learning, AI, and biological systems thinking (Goldstein, 2019; Michalski et al., 2013; Nevis et al., 2009), learning systems rely on multi-scalar, multidimensional feedback loops from the molecular to the institutional. They more resemble homeostatic systems (Goldstein, 2019) than open-loop systems, which typically follow rigid protocols and cannot or have difficulty adjusting to new information (Balashevich, 2009). Learning-systems approaches are not mechanistically reactive but structurally reflexive, dynamically and adaptively reshaping their own pathways as needed (Brinkerhoff et al., 2018).

Learning-systems approaches do not merely support better interventions for international development; they *become* the intervention. By integrating interdisciplinary knowledge with local contexts and embracing iterative methods, they create the conditions for communities to address root causes, not just symptoms (Brinkerhoff et al., 2018). They turn knowledge into an actionable, evolving foundation for decision-making. In doing so, they support not only resilience in the face of disruption but responsiveness to it. This responsiveness is crucial at every stage of the intervention value chain. The "last mile" problem—the difficulty of prompting end-recipients to take up and adapt a delivered solution—is mirrored upstream: basic researchers struggle to get attention by solution developers for overlooked problems; solution designers can fail to attract support for nonstandard solutions; project leads have to persuade funders to back methods that diverge from business-as-usual; and, above all, the diffusion of innovations must be relevant, accessible, and convenient for end-recipients (Bello-Bravo & Pittendrigh, 2023). A learning-systems approach discloses that the deeper challenge is not just delivery but mutual engagement, producing solutions that really solve people's real problems.

Ensuring solutions are co-defined and iteratively shaped by those who will live with them increases this likelihood, but this convergence is also where the right to knowledge becomes inseparable from learning-systems approaches. A learning system assumes that everyone has a right to understand, shape, and benefit from the information that affects their lives. This reiterates that knowledge must be accessible—in language, format, and relevance—especially to marginalized populations, including women, whose exclusion from public knowledge has historically been both structural and symbolic (Ellis, 1989; Lerner, 1986). This also means redefining what counts as learning: not just speedy changes of behavior but situated adaptation of solutions.

Material technology obviously helps, but social technologies like digital communities of practice, collaborative learning structures, and decentralized decision-making must be innovated and diffused as well. These resources are social architectures that empower and redistribute agency, embedding learning-change's adaptability into both sides of the infrastructure of development. In the current era, we also need systems that learn *quickly to adapt and support the SDGs*. The speed of a system's learning determines how effectively it can respond to emergencies, adapt strategies, or course-correct entirely (Dearden et al., 2007). Not all changes of behavior demand a course correction to a predetermined correct path; sometimes situations call for a change of destination in light of new information, a deeper realization about what the system should have been doing all along. Learning-systems approaches have more built-in tools to distinguish the needed course changes.

This requires institutions to override their own inertia. The longer a mistaken direction is followed, the more difficult it becomes to reverse, especially in multi-stakeholder environments where no one party can unilaterally pivot. Learning-systems approaches build in resistance to inertia not by top-down hierarchical control but through a disciplined openness: a constant habit not of just interpreting but inviting new information, questioning one's frames of reference, planning in the longer-term, and responding to evolving realities in real-time.

Learning-systems approaches operationalize the exercise of the right to knowledge, not as a passive entitlement but as an active, lived capacity. It makes visible the invisible labor of community-led solution-seeking, empowering local actors through collaboration, recognition, and adaptability. It recognizes the governance obligation of life-sustaining knowledge producers with respect to those who would use it. It does not just deliver information but adapts so that solutions will be taken up. In this sense, learning-systems approaches at critical links of the value chain are not merely means to better development and meeting the SDGs. They represent an expression of genuine development itself.

Arguably, all systems should be learning systems; the question then is: *how quickly can they learn?* The implications of adopting faster learning-systems approaches for achieving the SDGs are far-reaching and transformative. By integrating interdisciplinary knowledge and expertise, innovations and creative synergies cross sectors to generate more effective and multifaceted solutions (Ensor & Harvey, 2015), across multiple SDGs at once, improving health, poverty reduction, gender security, and resilience (Reynolds et al., 2018). Solutions become more sustainable, as communities are more likely to adopt and maintain practices that resonate with their cultural values and address their specific needs in each time, place, and circumstance (Rahman et al., 2023). The resultant empowerment not only can foster greater civic engagement, enhance transparency, and hold governments and institutions more accountable and transparent about their actions (Omanga, 2019) but also shift the demographic of what participations means—especially for women and marginalized people—even in the face of active opposition to it (Payumo et al., 2021).

Conclusion

Exercising the right to knowledge is essential for effectively advancing the SDGs. Knowledge access empowers communities and individuals to make informed decisions and adopt behavior changes that align intentions with outcomes at all scales—whether applying a postharvest loss prevention technique to support food security or ensuring that solution producers take up such innovations for broader implementation.

While the right to knowledge is widely recognized in matters of food, water, shelter, and health, it is equally vital in the pursuit of human dignity and gender security, not only the affirmation of women's right to access life-sustaining knowledge for themselves and their communities but a contribution to the deeper societal transformations that sustainability requires.

Knowledge access accelerates the spread of innovative social technologies and practices that promote food security, environmental stewardship, and collective well-being. It also strengthens educational initiatives by ensuring that even the most marginalized populations, particularly women, can exercise their right to knowledge for both individual empowerment and communal resilience. Upholding this right is fundamental to fostering inclusive, equitable development and ensuring that the benefits of pursuing the SDGs are truly shared by all. Learning-systems approaches enable this inclusion.

References

Agarin, T. (2020). The (not so) surprising longevity of identity politics: Contemporary challenges of the state-society compact in Central Eastern Europe. *East European Politics, 36*(2), 147–166. https://doi.org/10.1080/21599165.2020.1714598

Azcona, G., Bhatt, A., Davies, S. E., Harman, S., Smith, J., & Wenham, C. (2020). *Spotlight on gender, COVID-19 and the SDGs: Will the pandemic derail hard-won progress on gender equality?* UN Women.

Balashevich, N. V. (2009). Stabilization of linear systems by bounded controlling actions. *Journal of Automation and Information Sciences, 41*(5), 48–63. https://doi.org/10.1615/JAutomatInfScien.v41.i5.30

Bello-Bravo, J. (2023). "New wine in old bottles": Structures of feeling for a new way to see new wood products. *Forests, 14*(3), 524. https://doi.org/10.3390/f14030524

Bello-Bravo, J., & Lutomia, A. N. (2020). Supporting sustainability for a decent work and economic growth in Ghana. In W. Leal Filho, A. Azul, L. Brndli, A. Lange Salvia, & T. Wall (Eds.), *Decent work and economic growth* (Encyclopedia of the UN sustainable development goals) (pp. 1–9). Springer. https://doi.org/10.1007/978-3-319-71058-7_120-1

Bello-Bravo, J., & Pittendrigh, B. R. (2023). Getting in the loops: Using adaptive design methods for fall armyworm *Spodoptera frugiperda* (JE Smith) (Lepidoptera: Noctuidae) mitigation in Kenya, Nepal, and Bangladesh. *SageOpen, 13*(4), 21582440231220451. https://doi.org/10.1177/21582440231220451

Bello-Bravo, J., Medendorp, J. W., Lutomia, A. N., & Pittendrigh, B. R. (2023). *Gender, digitalization, and resilience in international development: Failing forward*. Routledge. https://doi.org/10.4324/9781003226383

Bhattacharya, R., & Bose, D. (2023). A review of the sustainable development goals to make headways through the COVID-19 pandemic era. *Environmental Progress & Sustainable Energy, 42*(4), e14093. https://doi.org/10.1002/ep.14093

Brinkerhoff, D. W., Wetterberg, A., & Wibbels, E. (2018). Distance, services, and citizen perceptions of the state in rural Africa. *Governance, 31*(1), 103–124. https://doi.org/10.1111/gove.12271

Brundtland, G., Khalid, M., Agnelli, S., Al-Athel, S., Chidzero, B., Fadika, L., Hauff, V., Lang, I., Shijun, M., & de Botero, M. M. (1987). *Report of the World Commission on Environment and Development: Our common future (O.U. Press, Ed.)*. World Commission on Environment and Development.

Daré, W., Barnaud, C., d'Aquino, P., Étienne, M., Fourage, C., & Souchère, V. (2014). The commodian stance: Interpersonal skills and expertise. In M. Étienne (Ed.), *Companion modelling: A participatory approach to support sustainable development* (pp. 41–67). Springer. https://doi.org/10.1007/978-94-017-8557-0_3

Davies, H., & Nutley, S. (2018). Organizations as learning systems. In D. Kernick & H. Bevan (Eds.), *Complexity and healthcare organization* (pp. 59–68). CRC Press.

Dearden, A., Light, A., Dray, S., Thomas, J., Best, M., Buckhalter, C., Greenblatt, D., Krishnan, G., & Sambasivan, N. (2007). *User centered design and international development CHI'07 extended abstracts on HUMAN factors in computing systems, San Jose, CA*. ACM Digital Library.

Ellis, K. F. (1989). *The contested castle: Gothic novels and the subversion of domestic ideology*. University of Illinois Press.

Ensor, J., & Harvey, B. (2015). Social learning and climate change adaptation: Evidence for international development practice. *Wiley Interdisciplinary Reviews: Climate Change, 6*(5), 509–522. https://doi.org/10.1002/wcc.348

Gates, B. (2020). Responding to Covid-19—A once-in-a-century pandemic? *New England Journal of Medicine, 382*(18), 1677–1679. https://doi.org/10.1056/NEJMp2003762

Goldstein, D. S. (2019). How does homeostasis happen? Integrative physiological, systems biological, and evolutionary perspectives. *American Journal of Physiology-Regulatory, Integrative and Comparative Physiology, 316*(4), R301–R317. https://doi.org/10.1152/ajpregu.00396.2018

Guha, J., & Chakrabarti, B. (2019). Achieving the sustainable development goals (SDGs) through decentralisation and the role of local governments: A systematic review. *Commonwealth Journal of Local Governance, 22*, 1–21. https://doi.org/10.3316/INFORMIT.896493850496702

Haji, M., & Himpel, F. (2024). Building resilience in food security: Sustainable strategies post-COVID-19. *Sustainability, 16*(3), 995. https://doi.org/10.3390/su16030995

Hollingworth, H. (1932). What is learning? *The Scientific Monthly, 35*(1), 63–65.

Lerner, G. (1986). *The creation of patriarchy*. Oxford University Press.

Lublin, D., & Bowler, S. (2018). Electoral systems and ethnic minority representation. In E. S. Herron, M. S. Shugart, & R. J. Pekkanen (Eds.), *The Oxford handbook of electoral systems* (pp. 159–174). Oxford Universiity Press.

Meuleman, L. (2022). A metagovernance approach to multilevel governance and vertical coordination for the SDGs. In A. Breuer, D. Malerba, S. Srigiri, & P. Balasubramanian (Eds.), *Governing the interlinkages between the SDGs* (pp. 71–89). Routledge.

Michalski, R. S., Carbonell, J. G., & Mitchell, T. M. (2013). *Machine learning: An artificial intelligence approach*. Springer Science & Business Media.

Mukarram, M. (2020). Impact of COVID-19 on the UN sustainable development goals (SDGs). *Strategic Analysis, 44*(3), 253–258. https://doi.org/10.1080/09700161.2020.1788363

Nevis, E. C., DiBella, A. J., & Gould, J. M. (2009). Understanding organizations as learning systems. In D. Smith (Ed.), *Knowledge, Groupware and the Internet* (pp. 43–64). Routledge.

Omanga, D. (2019). WhatsApp as 'digital publics': The Nakuru analysts and the evolution of participation in county governance in Kenya. *Journal of Eastern African Studies, 13*(1), 175–191. https://doi.org/10.1080/17531055.2018.1548211

Payumo, J., Bello-Bravo, J., & Pittendrigh, B. R. (2021). Demonstrating the nexus effects of online videos, research outputs, and investments to knowledge absorption using linguistically adapted animations. *European Journal of Applied Sciences, 9*(2), 283–301. https://doi.org/10.14738/aivp.92.10025

Rahman, M. F., Falzon, D., Robinson, S.-A., Kuhl, L., Westoby, R., Omukuti, J., Schipper, E. L. F., McNamara, K. E., Resurrección, B. P., & Mfitumukiza, D. (2023). Locally led adaptation: Promise, pitfalls, and possibilities. *Ambio, 52*(10), 1543–1557. https://doi.org/10.1007/s13280-023-01884-7

Rawls, J. (1971). *A theory of justice*. Harvard University Press.

Reynolds, M., Blackmore, C., Ison, R., Shah, R., & Wedlock, E. (2018). The role of systems thinking in the practice of implementing sustainable development goals. In W. Leal Filho (Ed.), *Handbook of sustainability science and research* (pp. 677–698). Springer.

Rodríguez-Domenech, M. Á., Bello-Bravo, J., & Pittendrigh, B. R. (2019). Scientific animations without Borders (SAWBO): An innovative strategy for promoting education for sustainable development. *Sustainability Science, 14*(4), 1105–1116. https://doi.org/10.1007/s11625-018-0624-8

Rodríguez-Domenech, M. Á., Bello-Bravo, J., Lutomia, A. N., Medendorp, J. W., & Pittendrigh, B. R. (2023). Digital education and community outreach for sustainable development in Africa. In W. Leal Filho, I. R. Abubakar, I. da Silva, R. Pretorius, & K. Tarabieh (Eds.), *SDGs in Africa and the Middle East Region: Implementing the UN sustainable development goals—Regional perspectives* (pp. 1–31). Springer. https://doi.org/10.1007/978-3-030-91260-4_65-1

Shanahan, G. (2024). Two routes to degeneration, two routes to utopia: The impure critical performativity of alternative organizing. *Organization, 32*(4), 13505084231223639. https://doi.org/10.1177/13505084231223639

Vargas, L., & Lee, P. (2018). Addressing communication and information poverty in the context of the sustainable development goals (SDGs). *Journal of Latin American Communication Research, 6*(1–2), 64–80.

Vargas, L., & Lee, P. (2023). Communication and information poverty in the context of the sustainable development goals (SDGs): A case for SDG 18—Communication for all. In J. Servaes & M. J. Yusha'u (Eds.), *SDG18 communication for all, volume 1: The missing link between SDGs and Global agendas* (pp. 25–59). Springer.

Winkler, I. T., & Williams, C. (2017). The sustainable development goals and human rights: A critical early review. *International Journal of Human Rights, 21*(8), 1023–1028. https://doi.org/10.1080/13642987.2017.1348695

Chapter 3
Empowering the Right to Knowledge: Lowering Transaction Costs for Educational Interventions and Communication Through Loosely Coupled Systems

Introduction

Coupling for Learning Systems

The previous chapter emphasizes the importance of feedback loops for reaching equilibrium in organic and non-organic systems. In this chapter, we explore further the continual adjustment and learning achieved by these learning systems through the concept of coupling. By coupling, we mean something subtly but importantly distinct from "connecting" or "linking." In the domain of biology, for example, Humberto Maturana and Francisco Varela (Maturana & Varela, 1987) characterize the relationship between an organism and its surroundings as structural coupling.

The concept of *coupling* emerged prominently in organizational theory through the work of Karl Weick (1976), who introduced the idea of "loose coupling" to describe systems where elements are interconnected but retain a degree of independence. Coupling refers to the strength and features of linkages between organizational components—such as units, processes, or goals—or between an organization and its external environment. Tightly coupled systems feature strong, direct, and immediate interdependence, where changes in one element rapidly affect others (Perrow, 1999). For example, in a tightly coupled organization, a policy change at the executive level might swiftly alter operational practices. In contrast, loosely coupled systems allow flexibility and autonomy, enabling subunits to adapt to local conditions without disrupting the whole (Weick, 1976).

Coupling frames the performance and success of each component in terms of the unit as a whole; in this sense, it reflects a conventional understanding of cooperation (Castañer & Oliveira, 2020). Coupling also involves a balanced measure of the involved parties' power dynamics and agency, usually associated with collaboration (Lutomia, 2019), even when there are steep hierarchical relationships in play (Lutomia et al., 2020). Thus, groups that are coupled can often act on behalf of, or

J. Bello-Bravo et al., *The Right to Knowledge*,
https://doi.org/10.1007/978-3-032-03360-4_3

with consideration for, partners in ways that are not strictly self-interested but pursue mutual benefits.

This associating again requires a closer distinction, since "connection" and "linking" can include cooperative and collaborative interactions between parties (Castañer & Oliveira, 2020; Scott & Davis, 2015). For institutional partnerships overall, this symbiotic mutualism is often a goal of the association. However, these arrangements usually do not prompt fundamental changes in the operations of either party or require prioritizing the partner's interests above their own, even when coordinating activities toward a common goal (Castañer & Oliveira, 2020; Hessels & Terjesen, 2010).

Likewise, coupling may not require such changes of behavior and priorities, but they often emerge as a consequence of the interaction. As such, there is an element of transformation that arises from coupling in contrast to connection and linking. Maturana and Varela (1987) usefully differentiate "organization" as the defining set of relationships and processes that constitute a system's identity, while "structure" refers to the physical components and their specific configurations within the system; in essence, "organization" defines what a system is, while "structure" determines how it is physically realized. In this way, not only will an organization maintain its identity despite many changes to the physical implementation of its structure, but it is also possible that past a certain point of such changes, the organization can actually change into something else. This alteration is the unique effect of coupling, where the identity of the organization becomes found not only in itself but also in and through the partner (Arango-Vasquez & Gentilin, 2021; Brusoni et al., 2001). In this sense, changes to structure tend to look like change in the conventional sense, while changes of organization may look like paradigm shifts (Depoux, 2009). It is these changes of organization (not just to structure) that we specifically ascribe to coupling as a critical aspect of learning systems.

The rationale and need for this organizational paradigm shift is both pragmatic and existential. For example, in response to changes in the working environment, what matters will not only be the solutions we generate as structural adaptations but also *how* we generate them, through shifts in organizational identity and practice (Bello-Bravo, 2023). One way to move toward changing this *how* can draw from the example of "edges" in organic and organizational relationships (Ferguson & Lovell, 2014; Vanni Accarigi et al., 2014).

Biologically, edges are zones of "maximal creativity" (LeVasseur, 2014) formed where distinct domains overlap—e.g., the shore, where land meets sea, or an estuary, where saltwater and freshwater converge. In these liminal spaces, new forms of life and behavior emerge that cannot be explained by either domain alone, such as amphibious or euryhaline species. Similarly, in human systems, collaboration brings together diverse perspectives to generate insights no individual could reach alone; similarly, cooperation combines distinct strengths to achieve outcomes that would be impossible independently. What distinguishes coupling is that these edge collaborations and co-operations lead to symbiotic transformations, mutually beneficial changes that reshape both participants. In each case, the meeting of two distinct systems gives rise to an unpredictable third, emergent system at their interface, reshaping both the boundary and the structure of the systems involved (Hallett & Hawbaker, 2021).

The common thread across these several examples is a creative emergence—of new species, identities, spaces, ideas, and behaviors at the boundaries—distinct from the domains they couple. As such, deliberately cultivating and leveraging these edges through coupling becomes a vital practice for learning systems at every level. Just as important is an organizational willingness to welcome and invite the changes coupling may bring, including their unexpected consequences and potential transformations. In this sense, edges and coupling offer a behavioral portrait of learning systems: organizations that open themselves to the possibility of paradigmatic shifts in both identity and structure, not just connection or linking to others.

Coupling Learning Systems for Development

The aim of this chapter is to describe learning-systems approaches for *coupling* the intellectual-research and resource-capacity aspects of the formal, non-formal, and informal development communities—particularly by lowering transactional cost barriers as gold-standard, scientifically based information moves from basic research through solution development, distribution, and end-user uptake. From experience in the field, effectively coupling the formal, non-formal, and informal sectors' diverse expertise and capabilities requires learning-systems to not only integrate these resources seamlessly but also build on any emergent insights that arise from their coupling (Johnson & Majewska, 2022). This integration yields more than just a pooling of intellectual capital and other resources, becoming more akin to the transformation of raw ingredients into a hearty stew. This integration, moreover, more adaptively and responsibly enables comprehensive and genuinely innovative solutions to support the right to knowledge and attaining the SDGs. By coupling resources, the development community will not merely "marshal" a broader range of capacities, skills, and knowledge cooperatively but will create the conditions for leveraging and exploring the resulting synergies—especially those interactions that lead to more effective and sustainable educational interventions as alternatives to business as usual.

However, even in its basic form, the organizational changes implied by learning-systems coupling are likely to encounter various forms of resistance, particularly when they require changes not only to established processes but also to broader institutional norms. Identifying metrics to evaluate practices at the interfaces with external organizations—especially grant-offering institutions—also presents challenges. What may appear as bureaucratic inertia from one perspective is, from another, at least partly a rational response to the specific operating conditions of an organization. As such, each organization will need to develop its own approach to change (Bello-Bravo et al., 2023). This is the "last mile" problem at the institutional level; thus, it is precisely to ease the logistical inconvenience of modifying institutional behaviors that decreasing transactional costs becomes relevant in this context.

From a technical standpoint, understanding the design and operational parameters of learning-systems approaches is fundamental to their successful implementation. This process includes identifying key features that encourage buy-in and participation—such as user-friendly interfaces, accessibility, and affordable mass-scalability—while addressing the technological, cultural, and socioeconomic barriers that have historically hindered adoption. By thoroughly examining these parameters, we can uncover pathways for installing systems that are not only robustly adaptable to both existing and emergent contexts but also capable of driving mass access to an actionable knowledge base.

Indeed, while the technical means to make life-sustaining knowledge available to the world's entire adult population are already well within reach, the potential remains limited without learning-systems approaches that reduce transaction costs at every stage—from buy-in around the creation of knowledge, through uptake of solution development and dissemination, to adaptation of solutions by end-recipients, and ultimately back to researchers, designers, and solution disseminators as feedback.

The most direct material illustration of these lowered transactional costs is the *decreasing* unit-cost to mass-scale the reach of animated video (Bello-Bravo et al., 2022a). This decrease is an entirely straightforward and direct result of the technical ease with which animated content can more cost-effectively adapted to new contexts than live-action video, often simply by translating the audio contents and re-dubbing them onto already-existing content—although revising the visual itself using preexisting asset libraries also greatly eases the costs of the task (Bello-Bravo, Medendorp, et al., 2023).

Overdubbing onto animated content also better avoids the distracting aspects that overdubbing onto live-action can cause, especially as unsynced dialogue (Szerszen, 2011). Overdubbing is also more widely usable than subtitling, not only because large segments of users may be non-print-literate (Fernández-Costales, 2021; Wu & Chen, 2021) but also because not all languages have print-forms for subtitling (Ong, 1982). Moreover, because animated video cost-effectively allows combining audio-overdubbing and subtitling (when feasible), leveraging both modalities significantly reduces the risk of the content being misunderstood, thanks to the multimodal (audio-visual) redundancy (Al-Fedaghi, 2012; Mayer, 1999, 2002). Most importantly, failing to provide life-sustaining knowledge in languages that individuals and communities speak is the most elementary denial of one's exercise of their right to knowledge (Chiblow & Meighan, 2022).

Transactional Costs

Understood pragmatically, "transactional costs" are the combined economic, social, ecological, and institutional costs involved in transferring knowledge from one domain to another across the entire research-development-distribution-feedback (RDDF) value chain—hence (1) from solution providers to end-recipients, (2) from solution designers to would-be distributors, (3) from solution researchers to would-be designers, and (4) from end-recipients to would-be new adaptations of the provided solution at some point along this value chain.

Lowered transactional costs can be understood by analogizing chemistry's notion of *catalysts*, which are agents that (1) enable or accelerate change (2) without themselves being consumed or fundamentally altered in the process (Bello-Bravo et al., 2023b). At all points along the RDDF value chain, learning-systems approaches can employ catalysts to decrease the costs of successfully "getting over the hump" of each step (Bello-Bravo & Pittendrigh, 2023) (see Fig. 3.1).

At the end of a value chain, the effects of catalysts can be seen in outcomes like strikingly high adaptation rates (89%) and knowledge retention (93%) of solutions delivered two years earlier (Bello-Bravo, Abbott, Mocumbe, Mazur, et al., 2019). But catalysts also act at the beginning of the value chain. For example, a chance encounter with a minority activist on Facebook created an edge: a liminal coupling where SAWBO's content could meet an urgent, unmet need. The activist's community had previously been denied access to COVID-19 prevention materials in their indigenous language due to discrimination by their national government. Within three days of the encounter, the activist and SAWBO collaborated to translate a preexisting COVID-19 prevention video into the activist's mother-tongue sharable with their community.

This outcome was not mere good luck but was possible because the preexisting COVID-19 prevention video served as a catalyst between SAWBO's infrastructure and the activist with reduced-enough transaction costs to quickly produce a local-language variant of the video that otherwise might not have existed. This chance meeting via Facebook was a novel interaction for SAWBO; as such, the coupling of this edge permanently altered both the activist's potential as a change agent and SAWBO's potential operation.

In cases of adaptation involving large numbers of participants, we can identify key parameters that either support, hinder, or have a neutral impact on people's ability to participate and exercise their right to knowledge. While numerous factors are known to facilitate the mass adoption of innovations, technologies, and systems (Rogers, 2003), the most consistent include ease of use, the ability of meeting diverse user needs, and perceived value (reinforced or not by the effectiveness of

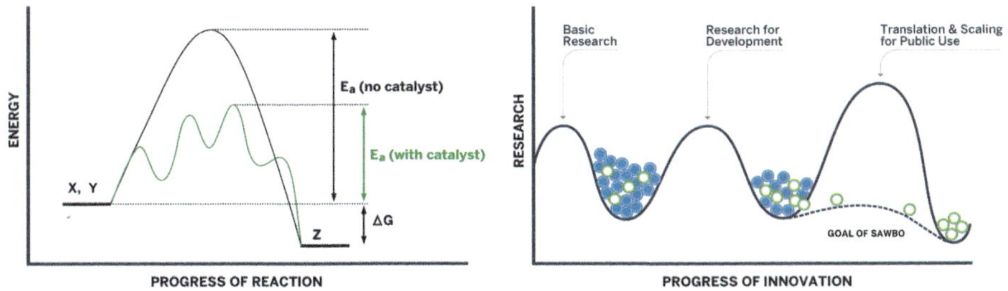

Fig. 3.1 Catalysts for change. Lowering energy costs in chemical reactions (on left); lowering transaction costs as a catalyst for uptake of innovations in SAWBO learning systems (on right). Reprinted with permission from Bello-Bravo et al. (2023a)

communication and community engagement strategies to promote it) (Rogers, 2003). Catalysts that lower transactional costs can significantly enhance these conventional facilitators, making adaptation more likely and more equitable.

One of the most salient catalysts for lowering transactional costs is *convenience*, i.e., when a course of action is not only available and culturally usable but also readily available to hand (Bhatt, 2017). In theories of action that invoke opportunity and motivation (Bos-Nehles et al., 2023), convenience can often seem to be folded somewhere into the picture by implication. However, even when motivation and opportunity appear strong enough to predict action, inconvenience can inhibit participation and information-seeking, suggesting it warrants being treated as a factor itself (Bello-Bravo et al., 2022b; Collier & Kimes, 2013; Connaway et al., 2011). The catalytic quality of convenience to lower transaction costs makes it invaluable for properly configured learning-systems approaches, as it lowers barriers to participation for people and facilitates a greater likelihood of their exercising their right to knowledge.

Technological Shifts, Access Barriers, and the Right to Knowledge

History exhibits hundreds of examples where shifts and innovations opened up previously inaccessible knowledge to large numbers of people. From the most famous example of Gutenberg's printing press to public libraries, the Internet, and mobile phones (Bello-Bravo et al., 2021), sociotechnological advancements have transformed access to information and knowledge—not only by enhancing accessibility and improving usability but also by reducing transactional costs to access that information and make it more convenient.

In the medieval era, accessing scholarly knowledge might require a pilgrimage to a distant location where the lone manuscript of a text was housed, with access restricted to elites possessing the proper clerical credentials. By the sixteenth century, Robert Estienne's (1572) celebrated *Thesaurus*—along with his notorious ex-employee's (Johannes Scapula) plagiarized and abbreviated but more affordable adaptation of it—made translating virtually the entirety of the extant classical Greek corpus into Latin accessible for anyone who could afford a copy. The equally compendious Liddell-Scott *Greek-English Lexicon* and its abridged Middle Liddell are now freely accessible online to anyone with a mobile phone browser through Tufts University's Perseus Digital Library.

Such historical instances give insights into how support for the right to knowledge can reach not just a broad audience but the entire adult population of the world. Implementing a learning-systems approach that draws on these past technological shifts and their conveniently increased access will further broaden access. By analyzing these dynamics, we can better understand the fitful progress of paradigm shifts and of system adaptation, ICT-upscaling, and design strategies that empower people's exercise of their right to knowledge.

In spite of these advantages, entrenched interests also resisted this progress while innovators and entrepreneurs advanced it in other contexts. But resistance can also be a simple disconnect. In a recent instance, the authors personally witnessed this situation. Although digital content had been delivered directly to extension agents in locally spoken indigenous languages for training farmers—and although low-cost technical equipment had been provided with training to use the equipment to overcome technological barriers—unfortunately, the content was never deployed because they "had never done it this way before." While such disconnects are not uncommon when new paradigms are proposed (Malhotra et al., 2021; Warrick, 2023), a more ideal solution would have allowed adjusting the situation so that not having done it before would not have been a barrier to delivery. Unhappily, this problem deprived not only the implementers of dramatically increased points of access but also the recipients of the life-sustaining knowledge.

In the most immediate sense, access to knowledge empowers individuals, communities, organizations, and coalitions to explore available technologies as potential solutions to local problems and to choose the options that best serve their best interests. Ideally, the desirability of a given technology should be self-evident, particularly when it delivers a minimally invasive, maximally effective short-term intervention with minimal long-term harm and significant short- and long-term benefits. Yet the issue is often more complex: the real question is sometimes not whether a given solution meets some criteria but whether even better alternatives might exist.

For instance, farmers in Burkina Faso—like many others—had turned to synthetic pesticides to combat crop losses from insects in a loosely regulated market. But they were not satisfied with a solution that, despite offering short-term gains, carried long-term environmental and health risks, especially when protective equipment was lacking or misused. Pesticides also could prove ineffective, as they were often counterfeited or adulterated. Recently, a change agent introduced the farmers to a non-synthetic pesticide made from neem tree seeds. While not as maximally effective as some synthetic pesticides (Aziz et al., 2013), it is environmentally non-harmful and could be sourced at limited-to-no cost to farmers after they locally planted neem and began producing their own pesticide (Bello-Bravo et al., 2023b). Not only is this outcome an excellent example of sustainability's triple-bottom-line for sustainable decision-making (Schweikert et al., 2018), it also illustrates how decisions around more and less desirable technologies can occur when adequate information is available.

Further, this question of desirability with respect to the introduction of new technologies and their effects on social interactions, community dynamics, and the environment is also not neutral: one must also ask *desirable for whom?* That some technologies foster greater collaboration and communication, while others exacerbate inequalities or instigate new problems, makes it necessary to critically assess who benefits from them—not only the positive and negative impacts of technologies but also ensuring they can be used in ways that empower the world's poorest and achieve the SDGs. It is unpalatable to contemplate, but for a host of the grosser inequalities we can diagnose as problems in societies, some sectors do not perceive these issues as problems at all, either because they are indifferent to and unaffected by them, or because they actually benefit from the social conditions that produce them. Usually framed as entrenched interests (Persha & Andersson, 2014;

Rajasekhar et al., 2018), these actors are typically well-positioned enough in terms of resources and power that they can actively resist or resiliently recover from changes of direction or policy that would otherwise reduce their influence. Worse, when such entrenched interests control the discourse of what can be deemed a desirable solution—and thus how one should "properly" go about solving a problem—this distortion affects other people's access to alternative (better) solutions. Nonetheless, learning-systems approaches demonstrate a capacity to deliver solutions even when local corruption would attempt to hinder delivery (Payumo et al., 2021). This recalls the activist who was able to access and help create translated COVID-19 prevention materials for their community.

These examples of bypassing culturally entrenched prejudices illustrate how the lowered transactional costs enabled by catalysts can counteract discriminatory and prejudicial forces that would otherwise diminish people's exercise of the right to knowledge. Whether this diminished access arises from deliberate self-interest or unintentional effects of the existing social order need not be resolved; the effects are real and consequential either way. Such discriminatory and prejudicial aspects result in the socially institutionalized, disproportionately disadvantageous denial of the right to knowledge for certain groups of people, solely based on their membership in various groups. In this sense, *discrimination* refers to the formal (*de jure*) expressions of otherwise prevailing (de facto) cultural *prejudices* that reproduce and reinforce diminished access. These pose both technical and social problems; that is, while technical solutions have addressed discrimination by repealing laws that legalize it (e.g., abolishing slavery) and by instituting new laws to protect against it (e.g., women's suffrage), they often do not eliminate their associated social prejudices (Alexander, 2012; Lerner, 1986).

Counter-intuitively, we can approach solving this problem of diminished access by following two key criteria: designing technical solutions that explicitly counteract social barriers and scale these solutions to tackle the most ubiquitous and culturally entrenched discriminations and prejudices, those obstacles based on gender (Agarwal, 2018; Lerner, 1986). The pragmatism and rationality of focusing on this large scale is partly due to the fact that mass-disseminating animated information has *decreased* unit costs to upscale (Bello-Bravo et al., 2022a) and because counteracting patterns of diminished access for nearly half of the world's adult population offers the greatest potential for increasing access.

A Time, A Place, and A Means

A technical approach to the social problem of women's diminished access to their right to knowledge must begin with securing the necessary time, space, and means to do so. In the twentieth century, civil disobedience movements offer powerful examples of how marginalized groups organized these conditions—often through strategic, centralized planning and highly visible mass mobilizations—to confront and dismantle legal discrimination, particularly through nonviolent protest (ICNC, 2000).

To the list of well-known factors that discourage, intimidate, or punish women for seeking to exercise their right to knowledge, mechanisms that disproportionately enable men's more convenient access can be added. Framed in terms of transactional costs, this convenience reflects systemic barriers that increase the difficulty for women to access knowledge. While overcoming such barriers also requires time, space, and technological means, its technical solution often differs from civil disobedience in its mode of action. It is ad hoc rather than planned, decentralized rather than coordinated, small-scale rather than mass mobilized—and often invisible, sometimes by necessity.

As a ready-to-hand catalyst for decreasing the transactional costs of accessing knowledge, mobile phone technologies are now unmatched, becoming in 2017 the most ubiquitous and technologically familiar form of digital access device globally (Bello-Bravo et al., 2021). However, while women and men both use mobile phone technologies, they do not do so in the same way or confront the same hindrances when doing so (Hafkin & Taggart, 2001; Huyer et al., 2005); for example, women's on-average lower incomes compared to men can make mobile phones prohibitively expensive (Wyche et al., 2016; Wyche & Olson, 2018). Women also can face familial and sexual safety issues not regularly experienced by men (Cumiskey & Brewster, 2012; Qushua et al., 2023; Wyche et al., 2016, 2019).

Other transactional costs include (1) disadvantageous class/educational norms that put mobile phone access prohibitively out of reach in terms of familiarity, time, equipment, and data plans (Bello-Bravo et al., 2022c; Gupta & Kiran, 2023; Mossberger et al., 2008) and (2) "geographic" norms—especially those rooted in colonial histories (Langa, 2022)—that neglect rural and remote regions (Langa & Bhatta, 2020).

One study in Bangladesh addressing gender inequities (SDG5) identified times, places, and means for lowering the higher transactional costs of moving available knowledge to women (Medendorp et al., 2022). It found that optimizing women's participation in the delivery of agricultural extension information—specifically in the places and at the times when women were more likely to be present to receive it—risked decreasing the number of people reached with the information overall (Medendorp et al., 2022; Reeves et al., 2023). However, subsequent research utilized strategic modeling of delivery to ensure both gains for women's participatory access and increased engagement overall (Reeves et al., 2023).

The first study measured attendance and other variables at "pop up" extension presentations offered at various times of day (e.g., morning, afternoon, and evening), different days of the week, and at several venue types (e.g., in religious institutions, markets, and at train crossings while pedestrians waited for trains to pass). From demographic data collected on the 131,073 farmers who participated in the 1070 pop-up trainings, we can determine which times and spaces were more likely to have Bangladeshi women present and willing to participate in ad hoc agricultural extension pop-ups (Medendorp et al., 2022).

For the population *overall* (including men and women and using *morning events* as a baseline for comparison), midday presentations had 33% lower odds of attendance, while late afternoon/evening presentations were 64% higher. In contrast, considering only women, they were more likely to be present at midday events and less likely to be present at late afternoon/evening events. Similarly (using *educational institutions* as a baseline venue for comparisons), women's much likelier participation at religious institutions and shops (355% and 177%, respectively) were accompanied by decreased likelihood of general population participation (35% and 26%, respectively). Tea stalls (114%) and farmer's houses (388%) were also popular venues for women; markets were not (69% less likely). Lastly, while overall attendance increased 16% more likely with male-led presentations, women were 47% less likely to participate in those pop-ups.

The second study used machine-supported decision-making to identify pop-up schedules that increased the odds of female participation by 82% while increasing participation overall by 16% (Reeves et al., 2023). This research shows that one can simultaneously aim to increase women's participation without a trade-off of decreased participation overall. As a technical solution, it provides a large-scale "end-around" on the social problem of prevailing prejudices that bar, discourage, or punish women's otherwise transactionally higher costs of exercising their right to knowledge. It mirrors the reduced transactional costs of the change activist's use of their mother-tongue variant of a COVID-19 prevention video to overcome the social problem of systemic discrimination.

Coupling, Decoupling, and Recoupling

We emphasized at the beginning of this chapter the importance of coupling to enable productive engagement across the information delivery value chain. Coupling creates fertile edges—dynamic zones of exchange where new ideas, methods, and approaches can be tested through symbiotic relationships between organizations. Whether these organizational levels are contextually formal, non-formal, or informal, it is unwise to confine interactions to a single type of partnership. Learning is most effective and organic when all three levels are activated, so long as the transaction costs of time, money, and effort are justified by the value of the exchange.

The successful delivery of timely content requires that all three levels of organizations be employed. However, there is always a risk, especially in international development, that coupling can lead to power imbalances, where international implementers dictate the scope or agenda of the relationship in ways that can be detrimental to the less-powerful or -influential partner (Lutomia et al., 2020). To be sensitive to these constraints, it is vitally important that engagements between organizations in highly diverse and complex contexts be loosely coupled so that they are also open to decoupling and recoupling, especially in times of dramatic shocks to the system (like COVID-19). One might argue that the 2025 deconstruction of the international development system and the tariff "reset" underway as we write is one of those occasions in which loosely coupled organizations will be in a better position to weather that storm.

In the field of international development, knowing when to decouple—especially in the area of organizational capacity development—can mean the difference between sustainable independence and chronic dependency (Incoom, 2024; UNDP, 2009). Decoupling, in organizational theory, refers to the gap between formal structure and actual practices (Boxenbaum & Jonsson, 2017). The current emphasis on finding local solutions is gaining traction, leading to accelerated decoupling in international collaborations (Witt et al., 2021). This expedience does not mean that collaborations are ending. On the contrary, international collaborations continue to grow (where they still are funded). Again, the uncertain and destabilizing systems deconstruction of the current moment in international development makes the ability to act outside of formal structures imperative.

The failures of international aid have been well documented (Easterly, 2006, 2007, 2009; Moyo, 2009). Developing engagements that honor the need to decouple for the well-being of partners is vital for the long-term success of engagement efforts. Jeffrey Pfeffer and Gerald Salancik's (Pfeffer & Salancik, 1978) resource dependence theory posits that organizations rely on external partners to access critical resources, but unequal control over these resources provokes power imbalances. In international contexts, power asymmetries are often exacerbated by differences in economic development, cultural norms, or regulatory environments (Cherbib et al., 2021; Faure et al., 2021; Gray et al., 2022; Hitt et al., 2000). Power asymmetries remain a leading cause of partnership instability and often lead to a lack of sustainability (Gulati & Sytch, 2007). If such collaborations are to succeed in the long-term, it is vital that there be intentional efforts to develop local capacity as well as knowing when to allow local leadership to take the initiative and when to disengage altogether. This option becomes only possible in loosely coupled organizations where decoupling is a possibility that allows for organizations on both ends of the international collaborations to act outside of their formal structures (Grant & Phene, 2022).

Although decoupling is an important phase for the growth and independence of international partners, especially those groups engaged in asymmetrical collaborations, it is equally important that organizations remain open to reengagement. Open platforms that allow for engagement and reengagement at any time—moving between formal structures and non-formal activities—are the only ones that will, in the long-term, permit constant growth and innovation in the knowledge space (Zámborský et al., 2023).

This latitude is due to the push and pull factors in the adoption of innovations (Chau & Tam, 2000). During the COVID-19 pandemic, for example, there was a dramatic rise in distance shopping through the use of digital technologies. The adoption of these technologies was driven both by the push factors (design, creation, and dissemination of new platforms) and the need to find alternatives to in-person shopping (Wang et al., 2021). In the push–pull framework, it is possible to push without pull, but adoption does not occur without pull (Moore & Benbasat, 1991). Organizational frameworks that allow for both push and pull in information dissemination become critical for achieving innovation adaptation. Maintaining a loosely coupled organization—as essential for facilitating push (availability of information) and pull (timely need for information)—is the best and perhaps only way to ensure long-term sustainability of technology adoption.

Returning to SAWBO's model (Fig. 3.2), the point of engagement with the potential client base is the open library platform that allows for push (making organizations aware of the informational tools available in the library) as well as pull (timely acquisition of the material based on local circumstances). Organizationally, SAWBO's platform is an intentionally open platform that allows for both formal structural engagement and non-formal, non-structured engagement.

Thus, SAWBO has entered into formal agreements with governmental and national agricultural NGO entities for the use of materials in their contexts and for cobranding of materials in local languages, but the platform always remains open for any person or group to download and use the material as is needed in their particular circumstances. Thus, anyone can pull from SAWBO's library and use the content when and where it is useful for them.

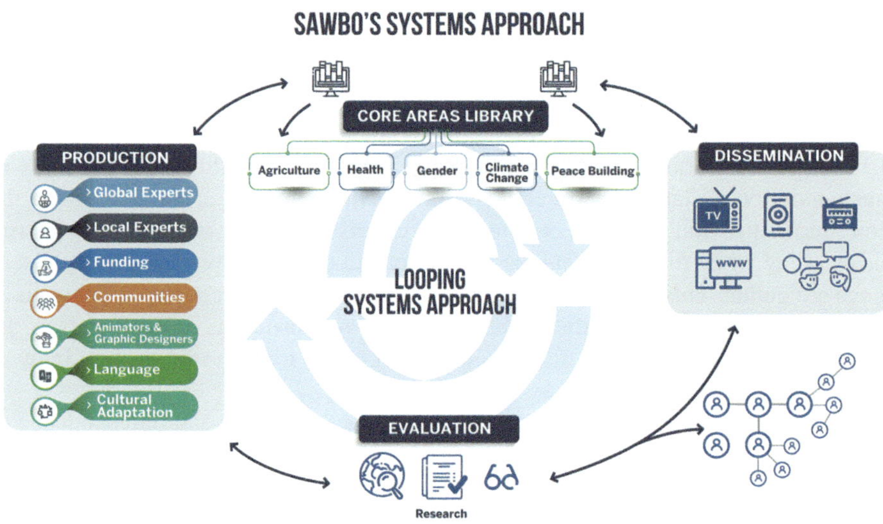

Fig. 3.2 SAWBO's learning-systems approach for reducing transaction costs and increasing accessibility: Production, core areas library, dissemination, and evaluation. (1) Once SAWBO animations are produced, (2) they are placed in an open-access, systems-based digital library for anyone to use, duplicate, and share as widely as possible. The library is accessible via YouTube (https://tinyurl.com/SAWBOYouTube), SAWBO's (2025) online library (https://sawbo-animations.org), and its Android-based app (https://tinyurl.com/SAWBOApp). SAWBO also provides an API that allows any other app to access and use its content. (3) This empowers SAWBO and any individual or group to download, disseminate, and share content across any system or platform they choose. (4) Finally, SAWBO and others can—and do—evaluate outcomes from using the content, contributing to public knowledge about how to most effectively leverage it. This cyclic learning-and-creation system results in an adaptive, openly accessible library of extension content, with tools for near-universal access and learning/research affordances that help expand collective understanding of how to increase the reach, engagement, use, and empowerment of such knowledge. This supports both the means and ends of the SDGs, and the exercise of the right to knowledge. Reprinted with permission from Bello-Bravo et al. (2023a)

For that reason, the consumer engagement part of the model remains open-ended. This feature of the platform has led to an unknown number of innovative and uses of SAWBO's material because there are no restrictions on use, other than for editing of the material to be done in collaboration with SAWBO to preserve the material's scientific integrity. For example, we are aware of groups using SAWBO's materials for fee-based agricultural advisory services, for a school gardens program that works with local teachers to provide non-formal extension services to school families, for formal government agricultural extension programs, and for informal WhatsApp communities of practice. These innovations will continue without limitations into the future because the knowledge base is freely accessible to all. Thus, SAWBO is and will remain a loosely coupled organization that allows for an unlimited number of uses, applications, and innovations through coupling, decoupling, and recoupling in a wide array of organizational contexts.

Nonetheless, to scale and reach millions, tens of millions, or billions of people, coupling needs to move toward the point where end users can access what they want, where they want, and then use and share it at their own convenience in times and places that make sense to them. This access is a coupling where SAWBO need not know of who is using the content.

To this end, SAWBO provides a range of platforms that facilitate widespread, flexible access to its educational materials. At the time of writing, SAWBO's YouTube channel hosts over 2100 videos, has more than 54,000 subscribers, and has accumulated 28+ million views from nearly every country in the world. In addition, SAWBO's online library (https://sawbo-animations.org) allows users to search for specific videos, select from available language options, and download content in multiple formats suitable for cell phones, computers, tablets, projectors, or television broadcast. For offline accessibility, SAWBO has distributed tens of thousands of credit-card-style flash drives containing curated video libraries for specific projects and regions—compact tools that allow recipients to store content in their wallets and transfer it when and where it is needed. Finally, SAWBO has developed a dedicated mobile app (https://tinyurl.com/SAWBOApp), refined over a decade, that enables users to access the full SAWBO video library, select preferred languages, and download content for use and sharing in-app or offline. Collectively, these platforms offer highly user-centered avenues for conveniently disseminating life-sustaining knowledge at scale.

Within four months of the SAWBO app's hard launch, over 60,000+ people worldwide had downloaded and used it, including users in remote and conflict-affected areas, such as Burundi, Zambia, the Democratic Republic of Congo, Tanzania, Niger, Nigeria, India, and Nepal. While SAWBO does not collect personal data, usage heat maps—based on users' voluntary permissions—indicate widespread, decentralized engagement. In one illustrative case, the app was accessed over fifty times in a single day in a remote Burundian village, without any direct contact between the user and SAWBO (Fig. 3.3). This is precisely the intended design: a scalable, self-sustaining system enabling individuals and organizations to access, download, and share life-sustaining R4D content in locally appropriate languages and formats. The model affirms the feasibility of extending access to such

Fig. 3.3 Sample daily use of the SAWBO mobile phone app in rural Burundi. In a single day, SAWBO's app was used over fifty times in a remote village in Burundi (center). The woman pictured on the right is a representative (not actual) user from the region, as the specific individuals or organizations using the app remain unknown to SAWBO unless use-location data is voluntarily shared. This anonymous and independent usage, without direct interaction with SAWBO, illustrates the platform functioning as intended by enabling local access to and dissemination of life-sustaining knowledge. (Source: Authors' own work. Image courtesy of Hilaire Muzibao)

content globally—across all of the world's approximately 7000 languages—as part the commitment to making life-sustaining knowledge accessible and the right to knowledge universally exercisable.

Conclusion

Integrating formal, non-formal, and informal development resources through well-designed, loosely coupled systems remains essential for improving the uptake of educational interventions. By understanding and addressing the parameters that shape how solutions are adapted, we can more intentionally engage with the opportunities emerging at the uncharted edges of technological change, drawing lessons from historical shifts and accounting for their social impacts. This approach enables the design of effective, inclusive knowledge dissemination systems that foster greater innovation and participation by local partners (Bello-Bravo et al., 2022b). Such systems provide a framework for ensuring fairness and accessibility, while minimizing transaction costs and maximizing both reach and impact.

To fully realize the potential of loosely coupled education systems in advancing societal progress and empowerment, the development community must actively support these emerging approaches. The use of animated videos in newly created spaces, where world-class research connects with locally driven initiatives, represents the kind of strategic edge that fosters innovation. SAWBO's intentional use of these edges through open systems, which distribute animated content to reduce transaction costs and overcome physical and social barriers, provides a clear example of how this approach can extend development efforts to communities that have historically been out of reach.

References

Agarwal, B. (2018). Gender equality, food security and the sustainable development goals. *Current Opinion in Environmental Sustainability, 34*, 26–32. https://doi.org/10.1016/j.cosust.2018.07.002

Alexander, M. (2012). *The new Jim Crow: Mass incarceration in the age of colorblindness*. The New Press.

Al-Fedaghi, S. (2012). A conceptual foundation for the Shannon-weaver model of communication. *International Journal of Soft Computing, 7*(1), 12–19. https://doi.org/10.3923/ijscomp.2012.12.19

Arango-Vasquez, L., & Gentilin, M. (2021). Organizational couplings: A literature review. *Innovar, 31*(79), 161–178. https://doi.org/10.15446/innovar.v31n79.91898

Aziz, M. A., Ahmad, M., Nasir, M. F., & Naeem, M. (2013). Efficacy of different neem (*Azadirachta indica*) products in comparison with imidacloprid against English grain aphid (*Sitobion avenae*) on wheat. *International Journal of Agriculture and Biology, 15*(2), 279–284.

Bello-Bravo, J. (2023). "New wine in old bottles": Structures of feeling for a new way to see new wood products. *Forests, 14*(3), 524. https://doi.org/10.3390/f14030524

Bello-Bravo, J., Abbott, E., Mocumbe, S., Mazur, R., Maria, R., & Pittendrigh, B. R. (2019). An 89% solution adoption rate at a two-year follow-up: Evaluating the effectiveness of an animated agricultural video approach. *Information Technology for Development, 26*(3), 577–590. https://doi.org/10.1080/02681102.2019.1697632

Bello-Bravo, J., & Pittendrigh, B. R. (2023). Getting in the loops: Using adaptive design methods for fall armyworm *Spodoptera frugiperda* (JE smith) (Lepidoptera: Noctuidae) mitigation in Kenya, Nepal, and Bangladesh. *SageOpen, 13*(4), 21582440231220451. https://doi.org/10.1177/21582440231220451

Bello-Bravo, J., Brooks, I., Lutomia, A. N., Bohonos, J. W., Medendorp, J. W., & Pittendrigh, B. R. (2021). Breaking out: The turning point in learning using mobile technology. *Heliyon, 7*(3), e06595. https://doi.org/10.1016/j.heliyon.2021.e06595

Bello-Bravo, J., Medendorp, J. W., Lutomia, A. N., Reeves, N. P., Tamò, M., & Pittendrigh, B. R. (2022a). Dramatically increased accessibility and decreased cost-per-person impacts are needed for scaling IPM in Africa. *Current Opinion in Insect Science, 54*, 100971. https://doi.org/10.1016/j.cois.2022.100971

Bello-Bravo, J., Medendorp, J. W., & Pittendrigh, B. R. (2022b). Just participation or just participation? A participatory justice model for more successful theory of change design, implementation, and solution uptake. *Heliyon, 8*(7), e09808. https://doi.org/10.1016/j.heliyon.2022.e09808

Bello-Bravo, J., Medendorp, J. W., Lutomia, A. N., & Pittendrigh, B. R. (2023). Contents, media, and genres for learning. In *Gender, digitalization, and resilience in international development: Failing forward* (1st ed., pp. 195–221). Routledge. https://doi.org/10.4324/9781003226383

Bello-Bravo, J., Muyodi, F. J., Nalwanga, R., Nakafeero, C. M., & Rutechura, F. R. (2022c). Time is not always money: A preliminary study on socially sustainable strategies banana xanthomonas wilts (BXW) mitigation efforts in Mbarara region Uganda. *Journal of Agricultural Extension and Rural Development, 14*(2), 90–101. https://doi.org/10.5897/JAERD2022.1319

Bello-Bravo, J., Rodríguez-Domenech, M. Á., & Pittendrigh, B. R. (2023). SDG1 in Europe: Micro-grants, poverty, and the big-picture future of sustainable development in a post-pandemic world. In M. L. De Lázaro Torres & R. De Miguel González (Eds.), *Sustainable development goals in Europe: A geographical approach* (pp. 3–20). Springer. https://doi.org/10.1007/978-3-031-21614-5_1

Bello-Bravo, J., Medendorp, J. W., Lutomia, A. N., & Pittendrigh, B. R. (2023a). *Gender, digitalization, and resilience in international development: Failing forward*. Routledge. https://doi.org/10.4324/9781003226383

Bello-Bravo, J., Medendorp, J. W., Lutomia, A. N., & Pittendrigh, B. R. (2023b). Wicked solutions for wicked problems: Catalysts and the WIDGET model. In *Gender, digitalization, and resilience in international development: Failing forward* (pp. 129–168). Routledge.

Bhatt, N. (2017). Relevance of universal human values and peace based education in democratic nations. *International Journal of Transformation in English & Education, 2*(2), 14–20.

Bos-Nehles, A., Townsend, K., Cafferkey, K., & Trullen, J. (2023). Examining the ability, motivation and opportunity (AMO) framework in HRM research: Conceptualization, measurement and interactions. *International Journal of Management Reviews, 25*(4), 725–739. https://doi.org/10.1111/ijmr.12332

Boxenbaum, E., & Jonsson, S. (2017). Isomorphism, diffusion and decoupling: Concept evolution and theoretical challenges. In R. Greenwood, R. E. Meyer, T. B. Lawrence, & C. Oliver (Eds.), *The Sage handbook of organizational institutionalism* (Vol. 2, pp. 77–101). Sage.

Brusoni, S., Prencipe, A., & Pavitt, K. (2001). Knowledge specialization, organizational coupling, and the boundaries of the firm: Why do firms know more than they make? *Administrative Science Quarterly, 46*(4), 597–621. https://doi.org/10.2307/3094825

Castañer, X., & Oliveira, N. (2020). Collaboration, coordination, and cooperation among organizations: Establishing the distinctive meanings of these terms through a systematic literature review. *Journal of Management, 46*(6), 965–1001. https://doi.org/10.1177/0149206320901565

Chau, P. Y., & Tam, K. Y. (2000). Organizational adoption of open systems: A 'technology-push, need-pull' perspective. *Information & Management, 37*(5), 229–239. https://doi.org/10.1016/S0378-7206(99)00050-6

Cherbib, J., Chebbi, H., Yahiaoui, D., Thrassou, A., & Sakka, G. (2021). Digital technologies and learning within asymmetric alliances: The role of collaborative context. *Journal of Business Research, 125*, 214–226. https://doi.org/10.1016/j.jbusres.2020.11.064

Chiblow, S., & Meighan, P. J. (2022). Language is land, land is language: The importance of indigenous languages. *Human Geography, 15*(2), 206–210. https://doi.org/10.1177/19427786211022899

Collier, J. E., & Kimes, S. E. (2013). Only if it is convenient: Understanding how convenience influences self-service technology evaluation. *Journal of Service Research, 16*(1), 39–51. https://doi.org/10.1177/1094670512458454

Connaway, L. S., Dickey, T. J., & Radford, M. L. (2011). "If it is too inconvenient I'm not going after it:" Convenience as a critical factor in information-seeking behaviors. *Library & Information Science Research, 33*(3), 179–190. https://doi.org/10.1016/j.lisr.2010.12.002

Cumiskey, K. M., & Brewster, K. (2012). Mobile phones or pepper spray? Imagined mobile intimacy as a weapon of self-defense for women. *Feminist Media Studies, 12*(4), 590–599. https://doi.org/10.1080/14680777.2012.741893

Depoux, D. (2009). New paradigms in organizational design. In A. Bausch & B. Schwenker (Eds.), *Handbook utility management* (pp. 105–133). Springer.

Easterly, W. (2006). Planners versus searchers in foreign aid. *Asian Development Review, 23*(02), 1–35. https://doi.org/10.1142/S0116110506500053

Easterly, W. (2007). Was development assistance a mistake? *American Economic Review, 97*(2), 328–332. https://doi.org/10.1257/aer.97.2.328

Easterly, W. (2009). Can the west save Africa? *Journal of Economic Literature, 47*(2), 373–447. https://doi.org/10.1257/jel.47.2.373

Estienne, H. (1572). *Thesaurus graecae linguae linguae, cui, conpingendo, additum est opus: glossaria duo…adutriusq. linguae cognitionem…perutilia. Item de atticae linguae seu dialecti idiomatis commentarium H. Stephanus* (Vol. Vol. 3). H. Estienne. https://books.google.com/books?id=RGSLLq3XK1sC

Faure, M. C., Munung, N. S., Ntusi, N. A., Pratt, B., & de Vries, J. (2021). Mapping experiences and perspectives of equity in international health collaborations: A scoping review. *International Journal for Equity in Health, 20*, 1–13. https://doi.org/10.1186/s12939-020-01350-w

Ferguson, R. S., & Lovell, S. T. (2014). Permaculture for agroecology: Design, movement, practice, and worldview. A review. *Agronomy for Sustainable Development, 34*(2), 251–274. https://doi.org/10.1007/s13593-013-0181-6

Fernández-Costales, A. (2021). Audiovisual translation in primary education. Students' perceptions of the didactic possibilities of subtitling and dubbing in foreign language learning. *Meta, 66*(2), 280–300. https://doi.org/10.7202/1083179ar

Grant, R., & Phene, A. (2022). The knowledge based view and global strategy: Past impact and future potential. *Global Strategy Journal, 12*(1), 3–30. https://doi.org/10.1002/gsj.1399

Gray, B., Purdy, J., & Ansari, S. (2022). Confronting power asymmetries in partnerships to address grand challenges. *Organization Theory, 3*(2), 26317877221098765. https://doi.org/10.1177/26317877221098765

Gulati, R., & Sytch, M. (2007). Dependence asymmetry and joint dependence in interorganizational relationships: Effects of embeddedness on a manufacturer's performance in procurement relationships. *Administrative Science Quarterly, 52*(1), 32–69. https://doi.org/10.2189/asqu.52.1.32

Gupta, M., & Kiran, R. (2023). Digital exclusion of women: A systematic review. *Global Knowledge, Memory and Communication, 74*(4), 938–957. https://doi.org/10.1108/GKMC-12-2022-0301

Hafkin, N. J., & Taggart, N. (2001). *Gender, information technology, and developing countries: An analytic study*. USAID—Office of Women in Development, Bureau for Global Programs, Field Support.

Hallett, T., & Hawbaker, A. (2021). The case for an inhabited institutionalism in organizational research: Interaction, coupling, and change reconsidered. *Theory and Society, 50*(1), 1–32. https://doi.org/10.1007/s11186-020-09412-2

Hessels, J., & Terjesen, S. (2010). Resource dependency and institutional theory perspectives on direct and indirect export choices. *Small Business Economics, 34*(2), 203–220. https://doi.org/10.1007/s11187-008-9156-4

Hitt, M. A., Dacin, M. T., Levitas, E., Arregle, J.-L., & Borza, A. (2000). Partner selection in emerging and developed market contexts: Resource-based and organizational learning perspectives. *Academy of Management Journal, 43*(3), 449–467. https://doi.org/10.5465/1556404

Huyer, S., Hafkin, N. J., Ertl, H., & Dryburgh, H. (2005). Women in the information society. In G. Sciadas (Ed.), *From the digital divide to digital opportunities: Measuring infostates for development* (pp. 135–195). Orbicom.

ICNC. (2000). *A force more powerful*. https://www.nonviolent-conflict.org/force-powerful-english/

Incoom, L. S. (2024). *Examining the impact of international aid dependency on governance and corruption in a African nations: A critical analysis of United Nations donations*. https://tinyurl.com/Incoom2024

Johnson, M., & Majewska, D. (2022). *Formal, non-formal, and informal learning: What are they, and how can we research them? Research report*. Cambridge University Press & Assessment.

Langa, N. (2022). *Historical patterns of regional, residential, and socioeconomic inequities in the utilization of maternal health care in Tanzania, East Africa* [Dissertation, University of Las Vegas]. Las Vegas, NV.

Langa, N., & Bhatta, T. (2020). The rural-urban divide in Tanzania: Residential context and socioeconomic inequalities in maternal health care utilization. *PLoS One, 15*(11), e0241746. https://doi.org/10.1371/journal.pone.0241746

Lerner, G. (1986). *The creation of patriarchy*. Oxford University Press.

LeVasseur, T. J. (2014). Permaculture. In P. D. Thompson & D. M. Kaplan (Eds.), *Encyclopedia of food and agricultural ethics* (pp. 1471–1473). Springer.

Lutomia, A. N. (2019). *A case study of successes and challenges in a scientific collaboration program based in the United States and Benin* [Dissertation, University of Illinois Urbana-Champaign]. Urbana, IL.

Lutomia, A. N., Bello-Bravo, J., Medendorp, J. W., & Pittendrigh, B. R. (2020). A positive project outcome: Lessons from a non-dominant government university-based program. *Interdisciplinary Journal of Partnership Studies, 7*(2), 3. https://doi.org/10.24926/ijps.v7i2.3482

Malhotra, N., Zietsma, C., Morris, T., & Smets, M. (2021). Handling resistance to change when societal and workplace logics conflict. *Administrative Science Quarterly, 66*(2), 475–520. https://doi.org/10.1177/0001839220962760

Maturana, H. R., & Varela, F. J. (1987). *The tree of knowledge: The biological roots of human understanding*. Shambhala.

Mayer, R. E. (1999). Multimedia aids to problem-solving transfer. *International Journal of Educational Research, 31*(7), 611–623.

Mayer, R. E. (2002). Multimedia learning. *Psychology of Learning and Motivation, 41*, 85–139. https://doi.org/10.1016/S0079-7421(02)80005-6

Medendorp, J. W., Reeves, N. P., Sal y Rosas Celi, V. G., Harun-ar-Rashid, M., Krupnik, T. J., Lutomia, A. N., Pittendrigh, B., & Bello-Bravo, J. (2022). Large-scale rollout of information and communication technology-enhanced extension training in Bangladesh demonstrates challenges and opportunities towards inclusive gender participation. *PLoS One, 17*(7), e0270662. https://doi.org/10.1371/journal.pone.0270662

Moore, G. C., & Benbasat, I. (1991). Development of an instrument to measure the perceptions of adopting an information technology innovation. *Information Systems Research, 2*(3), 192–222. https://doi.org/10.1287/isre.2.3.192

Mossberger, K., Tolbert, C. J., & McNeal, R. S. (2008). *Digital citizenship: The internet, society, and participation*. MIT Press. https://books.google.co.za/books?id=aH-5AAAAIAAJ

Moyo, D. (2009). *Dead aid: Why aid is not working and how there is a better way for Africa*. Farrar, Straus, and Giroux.

Ong, W. J. (1982). *Orality and literacy*. Methuen.

Payumo, J., Bello-Bravo, J., & Pittendrigh, B. R. (2021). Demonstrating the nexus effects of online videos, research outputs, and investments to knowledge absorption using linguistically adapted animations. *European Journal of Applied Sciences, 9*(2), 283–301. https://doi.org/10.14738/aivp.92.10025

Perrow, C. (1999). Organizing to reduce the vulnerabilities of complexity. *Journal of Contingencies and Crisis Management, 7*(3), 150–155. https://doi.org/10.1111/1468-5973.00108

Persha, L., & Andersson, K. (2014). Elite capture risk and mitigation in decentralized forest governance regimes. *Global Environmental Change, 24*, 265–276. https://doi.org/10.1016/j.gloenvcha.2013.12.005

Pfeffer, J., & Salancik, G. R. (1978). *The external control of organizations: A resource dependence perspective*. Harper & Row.

Qushua, N., Gillespie, A., Ramazan, D., Joergensen, S., Erskine, D., Poulton, C., Stark, L., & Seff, I. (2023). Danger zone or newfound freedoms: Exploring women and girls' experiences in the virtual space during COVID-19 in Iraq. *International Journal of Environmental Research and Public Health, 20*(4), 3400. https://doi.org/10.3390/ijerph20043400

Rajasekhar, D., Babu, M. D., & Manjula, R. (2018). Elite capture in decentralised institutions: A literature survey. In D. Rajasekhar, M. D. Babu, & R. Manjula (Eds.), *Decentralised governance, development programmes and elite capture* (pp. 19–29). Springer. https://doi.org/10.1007/978-981-13-1900-6_2

Reeves, N. P., Ramadan, A., Sal y Rosas Celi, V. G., Medendorp, J. W., Ar-Rashid, H., Krupnik, T. J., Lutomia, A. N., Bello-Bravo, J. M., & Pittendrigh, B. R. (2023). Machine-supported decision-making to improve agricultural training participation and gender inclusivity. *PLoS One, 18*(5), e0281428. https://doi.org/10.1371/journal.pone.0281428

Rogers, E. M. (2003). *Diffusion of innovations* (5th ed.). Free Press.

SAWBO. (2025). *Scientific animations without borders video library*. https://sawbo-animations.org/

Schweikert, A., Espinet, X., & Chinowsky, P. (2018). The triple bottom line: Bringing a sustainability framework to prioritize climate change investments for infrastructure planning. *Sustainability Science, 13*(2), 377–391. https://doi.org/10.1007/s11625-017-0431-7

Scott, W. R., & Davis, G. F. (2015). *Organizations and organizing: Rational, natural and open systems perspectives*. Routledge.

Szerszen, K. A. (2011). *The audio/visual mismatch and the uncanny valley: An investigation using a mismatch in the human realism of facial and vocal aspects of stimuli* [Thesis, Indiana University]. Bloomington, IN.

UNDP. (2009). *Capacity development: A UNDP primer*. UNDP.

Vanni Accarigi, I., Crosby, A., & Lorber-Kasunic, J. (2014). Value the edge: Permaculture as counterculture in Australia. *M/C Journal, 17*(6). https://doi.org/10.5204/mcj.915

Wang, X., Wong, Y. D., Chen, T., & Yuen, K. F. (2021). Adoption of shopper-facing technologies under social distancing: A conceptualisation and an interplay between task-technology fit and technology trust. *Computers in Human Behavior, 124*, 106900. https://doi.org/10.1016/j.chb.2021.106900

Warrick, D. (2023). Revisiting resistance to change and how to manage it: What has been learned and what organizations need to do. *Business Horizons, 66*(4), 433–441. https://doi.org/10.1016/j.bushor.2022.09.001

Weick, K. E. (1976). Educational organizations as loosely coupled systems. *Administrative Science Quarterly, 21*(1), 1–19.

Witt, M. A., Li, P. P., Välikangas, L., & Lewin, A. Y. (2021). De-globalization and decoupling: Game changing consequences? *Management and Organization Review, 17*(1), 6–15. https://doi.org/10.1017/mor.2021.9

Wu, Z., & Chen, Z. (2021). A systematic review of experimental research in audiovisual translation 1992–2020. *Translation, Cognition & Behavior, 4*(2), 281–304. https://doi.org/10.1075/tcb.00056.wu

Wyche, S., & Olson, J. (2018). Kenyan women's rural realities, mobile internet access, and "Africa Rising." *Information Technologies & International Development, 14*, 33–47.

Wyche, S., Simiyu, N., & Othieno, M. E. (2016). Mobile phones as amplifiers of social inequality among rural Kenyan women. *ACM Transactions on Computer-Human Interaction (TOCHI), 23*(3), 1–19. https://doi.org/10.1145/2911982

Wyche, S., Simiyu, N., & Othieno, M. E. (2019). Understanding women's mobile phone use in rural Kenya: An affordance-based approach. *Mobile Media & Communication, 7*(1), 94–110. https://doi.org/10.1177/2050157918776684

Zámborský, P., Ingršt, I., & Bhandari, K. R. (2023). Knowledge creation capability under different innovation-investment motives abroad: The knowledge-based view of international innovation management. *Technovation, 127*, 102829. https://doi.org/10.1016/j.technovation.2023.102829

Chapter 4
Knowledge Engagement and ICT-Scaling: Gold-Standard Knowledge for Global, National, and Community Change

The Role of Information and Communication Technologies (ICTs) in Enhancing Animation Use for Learning

In recent years, digital ICTs have extensively impacted the use of animations in educational contexts, especially around accessibility and engagement. This development builds on the mobile phone becoming, since 2017, the primary device type for accessing digital information (Bello-Bravo et al., 2021), even in places where digital infrastructures and Internet access, data plans and access costs, and social factors hinder their use (Bello-Bravo & Pittendrigh, 2012; Gumucio et al., 2019; Huyer, 2016; Payumo et al., 2021).

This chapter explores how animated video media, accessed via mobile phones and other video-capable devices, has become a critical pathway for mass-scaling access to vital learning content. These digital platforms have the potential to empower audiences who, under traditional extension systems, might otherwise remain unreached. While limitations remain—many of which are addressed later in the chapter—we begin here by outlining the key effects of this system. As discussed in the previous chapter, providing animated content through digital platforms in a loosely coupled arrangement with end users leverages the widespread availability and accessibility of mobile technology.

Although animated content can and is used in traditional, in-person extension systems, digital ICTs significantly enhance the mass dissemination of such educational materials. They help ensure that content reaches a broad and diverse audience by combining push strategies (to raise awareness of the content's availability) with pull mechanisms that rely on end users' interest or need for the information. This dual approach not only enriches the overall learning experience but also strengthens the credibility of the information, as users engage with content in familiar formats and languages. ICTs, therefore, serve as critical tools for bridging educational gaps and promoting inclusive learning by offering alternative pathways to access

J. Bello-Bravo et al., *The Right to Knowledge*,
https://doi.org/10.1007/978-3-032-03360-4_4

information that bypass traditional barriers. This model demonstrates a strong potential for improving the adoption of solutions, particularly by those typically excluded from formal systems (Bello-Bravo et al., 2020a). While the focus here is on empowering end users and addressing the problem of the "last mile," the approach also fosters greater buy-in throughout the research–development–dissemination value chain, enhancing the uptake of innovations by both producers and those responsible for delivering them.

Dramatically Reducing Costs for Adapting Educational Methods to African Contexts

Sustainability—originally formulated in the *Brundtland Report* (Brundtland et al., 1987)—mandates giving "overriding priority" to meeting the needs of the world's poor. By this standard, Africa is a central focus, as it is the continent with the lowest nominal GDP globally (see Fig. 4.1). Accordingly, to design educational materials and methods with Africa in mind serves a dual purpose: it directly addresses the most urgent needs while also ensuring that these approaches remain adaptable and effective in less constrained contexts.

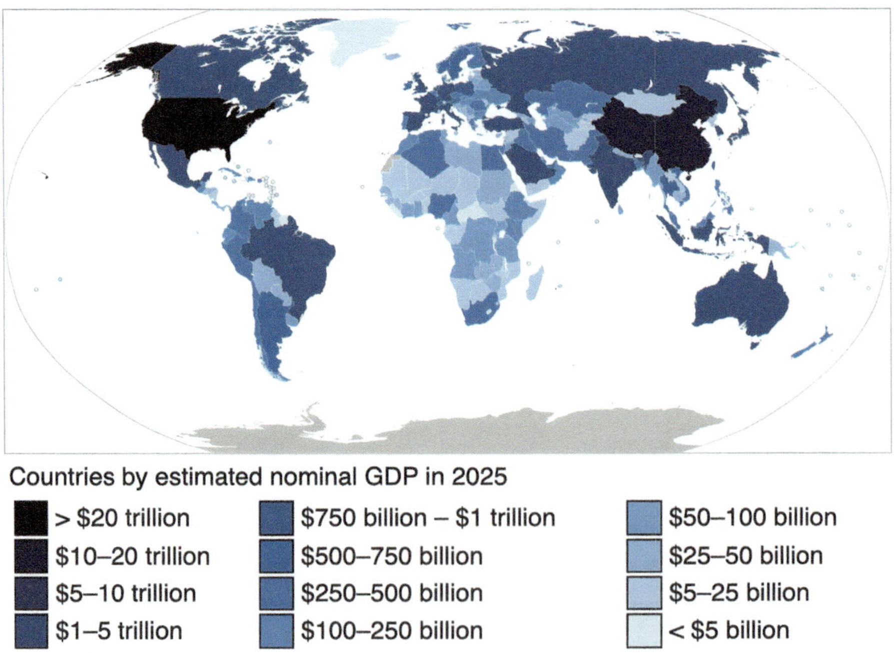

Fig. 4.1 Countries by estimated nominal GDP. This world map uses color to indicate the nominal gross domestic product (GDP) ranges of each country. (Reprinted from Anders Feder (2024), CC BY-SA 4.0)

Digital infrastructures are expanding across Africa but are still limited by access and cost, preventing universal participation (Corrigan, 2020; Nchake & Shuaibu, 2022). Radio and television continue to play important roles in information campaigns to remote locations in countries (Attuh & Kankam, 2024; Iacoella et al., 2022; Komodromos, 2024; Williams & Tkach, 2022). Other channels for distributing information, especially extension services, tend to be underfunded and short of personnel—a situation not likely to improve on its own given international development's current deconstruction. Agent-to-farmer ratios in sub-Saharan Africa range between 1:3000 and 1:15,000 (AGRA, 2020; Davis et al., 2019). This ratio is not for want of good intentions; limited national budgets and competing priorities curtail available monies for this key service. Moreover, the unparalleled diversity of languages and cultures in Africa complicate educational campaigns, since delivering content in a single language (even a national one) can leave minority language communities unreached (Reeves et al., 2024).

People who are most isolated linguistically, particularly those speakers with little or no print-literacy in widely spoken national languages, are often the most overlooked, underserved, and excluded from access to vital information. Animated videos offer a practical solution to bridge this gap. With low production costs and high adaptability, they can be modified to suit virtually any linguistic and cultural context, making it possible to deliver gold-standard, life-sustaining knowledge to even the most geographically remote or marginalized communities, including those in Africa. This adaptability supports the SDGs and exercise of the right to knowledge by providing a foundation for mass-scalable, non-exclusive distribution systems that can reach everyone with information they seek to improve their lives.

For any e-extension system to be broadly applicable, strategies to increase accessibility and reduce costs are indispensable (Bello-Bravo et al., 2022). Questions then arise which cost parameter(s) are critical overall and best enhance cost-effectiveness for accessibility. For example, decision-making is needed for placing content into rare or minority languages, synergizing the benefits of doing so with the seeming cost of smaller numbers positively impacted (Reeves et al., 2024). SAWBO addresses this issue by collaborating with global experts and animators to develop animations at a set production cost. Once created, the animation can then be translated into other languages at a much lower cost, often around two orders of magnitude less. To be sustainable, the original content must reach a large enough audience to afford the initial investment and the reduced costs of translating and localizing content for minority and rare languages—some of which may be threatened with extinction (Bello-Bravo et al., 2022).

Other disciplines can offer valuable insights into how to model such systems—for example, resistance management in transgenic crops engineered to control insect pests, such as *Bt* corn. Although the first transgenic plant was developed in 1983 (De Framond et al., 1983), *Bt* corn gained widespread commercial use in the United States by 1996. *Bt* corn contains genes from the bacterium *Bacillus thuringiensis* (hence *Bt*) that produce proteins toxic to specific insects. When ingested, these proteins disrupt the insect's gut cells, leading to death. However, over time, insect populations can evolve resistance, selecting for pests that are unaffected by the *Bt* proteins and eventually rendering the crop protection ineffective. To delay this process, scientists developed a strategy known as "refuges," in which a portion of the field is planted with non-*Bt* crops. This allows non-resistant insects to thrive and interbreed with any resistant individuals, thereby maintaining a population that remains susceptible to the *Bt* proteins. The refuge strategy effectively slows the evolution of resistance and extends the functional lifespan of the *Bt* gene.

The solution was developed through a computational model borrowed from population genetics called the Hardy-Weinberg principle (Chen, 2010). Building on these early models, resistance management emerged to study factors influencing resistance while determining optimal ratios of *Bt* to non-*Bt* plants to preserve insect population vulnerability. As the science developed, researchers deepened their understanding of the dynamic interactions between insect populations and *Bt*-resistant crops, allowing for more precise strategies to prolong the effectiveness of *Bt* genes, potentially indefinitely. By identifying key factors and modeling their parameters, researchers could calculate optimal crop configurations, refuge sizes, and population thresholds to manage resistance over the long term.

The ability to model future outcomes based on factors and parameters is not unique to plant and insect sciences. Many disciplines model probabilistic future outcomes and develop strategies for effectiveness. The discipline of scalable e-extension can also benefit from such computational models. To achieve this advantage requires identifying key parameters influencing the creation–dissemination–access–learning–adaptation value chain and designing scaling models to optimize each segment of the chain. A decision-tree model (Fig. 4.2) can include specific parametric factors into which values can be inserted to model the cost-effectiveness of moving forward with creation of a given animation.

These or similar models are then applied to other value-chain segments so that as time goes on, the modeling becomes more refined and useful. One could ask the question: *at what point will the investment in a minority language translation pay for itself in terms of increased benefits to those who use the technologies or educational content*?

Modeling Cost-Effective Extension Systems

As statisticians George Box and Norman Draper (1987) famously observed, "All models are wrong; the practical question is how wrong do they have to be to not be useful" (p. 74). In this context, "useful" refers to a model's ability to support meaningful results. Scalable e-extension systems are precisely the kind of space where such modeling is needed, combining theoretical frameworks with empirical testing through both on-the-ground and online data collection, followed by repeated refinement.

As part of ongoing research into e-learning platforms, SAWBO is collaborating with modelers trained in control theory, an engineering discipline focused on maintaining system equilibrium in response to internal and external stresses (Ortega &

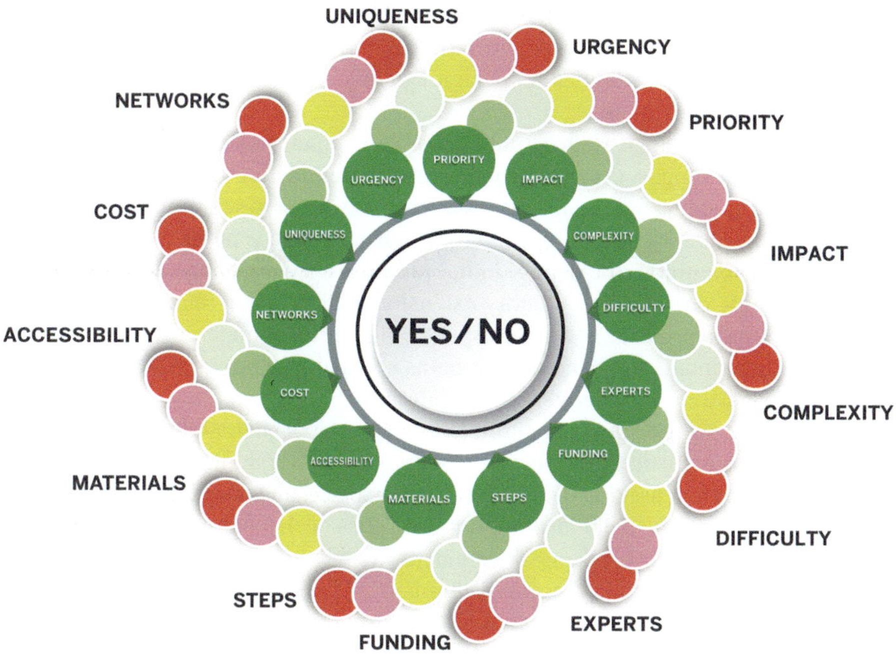

Fig. 4.2 Sample decision-tree model for content creation. Used by SAWBO to determine and prioritize content creation, this type of model incorporates weighted parametric factors, such as uniqueness, urgency, priority, and steps, into a structured decision process. Quantitative values weight each node to assess the overall cost-effectiveness of producing a given animation, helping guide strategic choices around which content to create and deploy. Reprinted with permission from Bello-Bravo et al. (2023)

Lin, 2004). Commonly applied in aeronautics, this theory provided SAWBO a framework to evaluate variables affecting scalable e-extension systems globally, but especially in Global South contexts. Together with economists, the team is developing models to assess the production and deployment costs of animated videos for digital extension, including both content creation and the cost of language adaptations. The model was used to calculate break-even points and potential returns on investment for these interventions.

Assuming a 30-year time horizon, the model will show how new agricultural e-learning content could reach its break-even point with just over 100 users adopting a featured technology (unpublished). For language translations of existing content, that number drops to fewer than ten users. In other words, animations created for widely spoken African languages—like Swahili, Hausa, or Yoruba—can become cost-effective even with relatively low uptake, while also supporting translation into minority languages despite small speaker bases.

Preliminary research also suggests that cost of adopting a specific innovation—whether $1 or $10—can have little effect on return on investment. This suggests that low-cost (and even medium-cost) innovations, such as the use of jerrycans (low-cost) or similar items (hermetically sealed bags; medium-cost), remain viable targets for animation content, even when they require small financial commitments. This insight opens opportunities for designing content around a wide range of practical, affordable innovations.

The model also suggests that timelines shorter than typical ones development projects can still support cost-effective content dissemination. For example, within a 3-year investment window, the break-even point for a new animation could be approximately 7593 users, and just 130 for a new language version. Such figures are readily achievable through existing digital platforms. This amount underscores the value of creating durable, high-quality extension materials that retain relevance and usability beyond a project's formal lifespan. Unlike printed materials, which require ongoing reinvestment, digital content remains accessible at minimal additional cost, supporting long-term knowledge dissemination.

While these specific figures are ultimately model-dependent, their broader implication is clear: linguistic adaptation of essential agricultural knowledge may be far more affordable than previously assumed. Historically, a lack of language diversity in learning materials (usually also in print form) has been a major barrier for conveniently accessing and exercising the right to knowledge. This research holds the promise that the transactional costs of that barrier will be lowered enough via the use of animated videos distributed through loosely coupled e-learning platforms. This will represent not just a technical improvement but a strategic breakthrough, one that lays a foundation for making the right to knowledge truly universal.

Digital Versus Traditional Extension

Traditional extension services typically involve trained technical advisors traveling to communities or hosting centralized events. These efforts often include demonstrations, printed materials, lodging, per diems, and transportation, all of which raise per-person costs significantly. Estimates suggest in-person extension services cost around $35 per recipient, though these figures often omit broader administrative overhead. Regardless of the exact figure, studies consistently show that digital methods are far more cost-efficient (Bernard et al., 2019; Khatri et al., 2024).

In contrast, ICT-based systems rely on scalable platforms such as YouTube, WhatsApp, Facebook, and Telegram to distribute content. These platforms reduce the need for a specialized technical workforce to deliver information. Instead, much of the expertise is embedded directly in the animated video content, which is scientifically validated and accessible even to low- and non-literate audiences. This is not a proposal for digital-only extension; the "human" element remains a crucial part of extension, not only because people can prefer face-to-face rather than computerized or remote interactions but also because facilitated discussions about video content show better learning outcomes (Bello-Bravo et al., 2018c; Tata & McNamara, 2018). Digital approaches should complement existing extension efforts and outright provide the service where existing extension systems cannot answer the demand.

With such cost-effective methods of content dissemination, it is now possible to imagine that R4D innovation distribution might reach a tipping point to become "common knowledge." Precisely this result seems to have occurred in parts of Kenya, where SAWBO's jerrycan video for hermetically storing beans postharvest had been widely distributed on social media. Years later, hearing the mother of a colleague explaining the video's technique, we asked when she had seen it, but she said didn't need to: "Everyone knows how to do it—it's common knowledge." This is precisely an ideal outcome—that life-sustaining knowledge becomes widespread enough to become common knowledge with no need for technical intervention to propagate the knowledge.

While our current models (unpublished) have yet to exactly zero in on the complex relationship between those who are exposed (as reach and engagement) to new knowledge content and those who use and adapt it, early results suggest that reach and engagement costs can be reduced to as low as \$0.01–\$0.10 per person. Assuming the \$0.01–\$0.10 range, even with only 20–25% uptake, the effective cost per adopter would remain \$0.04–\$0.50, a significant improvement over traditional methods. The model also suggests that returns on investment rise sharply as exposure costs fall. At \$10 per exposure, returns are modest regardless of reach; at \$5 or \$1, gains are incremental. However, at \$0.50 per exposure, returns on investment increase substantially and grow exponentially at \$0.10 and below. Driving these costs to pennies (or fractions of pennies) per person reached will yield a considerable return on investment.

Relatedly, using AI to target high-interest viewers via YouTube, we have achieved full-video views for as little as 3.7 cents and partial views (30% watched) for just 1 cent. Full views indicate reach and engagement without a guarantee of uptake but do indicate a strong interest that likely correlates with higher likelihood of use. The SAWBO app also lowers the cost of acquiring new users to 5 cents each. These viewers can download videos and unlimitedly share them offline or through digital social platforms like WhatsApp. This structure enables a scalable system where a single 5-cent investment in an app user can lead to multiple additional viewings, further reducing the per-person cost of delivering essential knowledge.

Commonly used social media platforms will differ across countries and change over time. SAWBO's app is designed to adapt to this variability by allowing users to easily access relevant videos and share them across whichever analog and social media channels make the most sense for their context. Videos stored on devices become shareable offline anywhere and wherever it is convenient. The core point is not the app itself, but the broader opportunity it represents: a scalable system for delivering knowledge to newly informed recipients at costs below the \$0.01–\$0.10 per person per video.

Moreover, our experience in Africa suggests that the movement of technologies and techniques from urban to rural settings is more common than more industrialized societies because the social divide between the rural and the urban is less sharp. In Africa—as also in many regions under development—urban residents, who have more access to technology than their rural counterparts, are often first-generation city dwellers with strong ties to the rural landscape; in other words, they still have strong roots in the countryside. Even multi-generational urban families frequently either maintain family linkages with rural areas, which they still visit regularly, especially around holidays, or are still steeped in the agricultural milieu of their cultures and aspire to, or have achieved, rural land ownership and engage in hobby or commercial farming. Under these circumstances, the flow of information from higher-tech urban areas to lower-tech rural areas is relatively unimpeded.

Realizing the right to knowledge as part of attaining the SDGs requires adopting new methods and models not just for extending the reach and engagement (delivering) extension content at far lower costs than traditional development approaches but the likelihood of uptake and adaptation as well. Many (once) well-funded NGOs and government extension systems rely on in-person delivery by trained technicians that are (now prohibitively) expensive, limited in reach, and often ineffective, because they depend on push rather than pull dynamics.

In contrast, the major advantage of well-designed ICT resources is not just their low cost but their constant availability—accessible to anyone, at any time, in any location. This dissemination, moreover, is driven primarily by pull factors, which are essential for bridging the gap from reach and engagement to adaptation and actual use. In-person participation often pulls in "gatekeepers of knowledge" who may pass on little to none of what they learn (see Rajasekhar et al., 2018 on elite capture). This flaw is neither to dismiss trainings and demonstrations to raise awareness and increase the pull factors that drive use and adaptation, nor the large number of people sincerely involved in such events who seek positive impacts in their communities. Nor are we disregarding the problematic aspects of social media and gendered access issues around digital ICTs (Adavant, 2021; Ahad & Lim, 2014; Lutomia et al., 2024; Omanga, 2019). It is only that SDG efforts will no longer have to rely exclusively on extremely high-cost, reach- and engagement-limited approaches, holding all else equal. We have the opportunity now to fully transition to models where we can reach and engage mass-subsets of populations with a broad base of life-sustaining and potentially "common knowledge"—especially with those otherwise without access to the "have" side of the digital divide.

Animation as an Ideal Medium for Knowledge Transfer

In other chapters of this book, we distinguish *information* from *knowledge* and formal, non-formal, and informal education as part of a learning-systems approach for mass-scaling information to everyone. Here, we use these terms in a less technical way, largely for the sake of the exposition's clarity and comprehensibility.

Both in popular and academic practice, disseminating or transferring "information" or "knowledge" can be roughly understood as formal, non-formal, and informal "education" or "learning" (both digital and analog), where the extent of that dissemination as "distribution" or "diffusion" can be measured in terms of "reach" (the potential exposure of knowledge through channels), "engagement" (the actual encounter with that content by someone), and the "uptake," "buy-in," "use," or "adoption" of that knowledge (as its adaptation and utilization by someone). In this general way, "empowerment" then means a facilitation of changes to habitual behavior as learning that results from adapting and improving one's life and community with such knowledge. Moreover, SAWBO's goal—consistent with the SDGs—is to positively transform people's lives with solutions that not only *work* but also yield changes to habitual behavior providing "an expansion in people's ability to make strategic life choices, in a context where this ability was previously denied to them" (Kabeer, 1999, p. 437). This result is the intended outcome of a person's exercise of their human right to knowledge—a healthier, safer, more dignified, and fulfilling existence.

A moment's reflection might lead some to doubt whether it is truly possible to reach, engage, and empower *everyone* with life-sustaining knowledge, even though that is the SDGs' stated aspiration. But we must not plan to leave anyone behind, whether by oversight or design. This kind of skepticism is not just premature; it assumes in advance that knowledge cannot reach certain populations and, worse, rationalizes efforts that treat exclusion as inevitable. Whether the doubt seems reasonable or not, the effect is the same: it guarantees that the same groups who have always been overlooked or deliberately excluded will once again be denied access to gold-standard, scientifically grounded knowledge and solutions (Bello-Bravo et al., 2018b). The dramatically lowered transactional costs, which make possible mass-scaling knowledge to the widest possible demographic, afford conditions for reaching "common knowledge" tipping points. Under such conditions, the right to knowledge becomes meaningful, since a right that cannot actually be exercised is no right at all.

The medium of SAWBO's animated content is crucial to its reach, engagement, and empowerment. Animation as a genre has been shown to be at least as effective as live-action video in transferring knowledge (Smith et al., 2012) and more cost-effective to produce, especially when scientific experts are engaged in content creation (Bernard et al., 2019). Crucially, animations can visually abstract and simplify complex procedures without sacrificing informational integrity, also making them easier to remember (Lowe, 2001, 2003). This ease is especially vital when learners must reproduce a protocol to achieve scientifically validated outcomes (Bello-Bravo et al., 2018a).

Teaching scientifically vetted information through animation also ensures consistency and trustworthiness (Bello-Bravo et al., 2020a; Mocumbe, 2016) and measures higher knowledge gains compared to other forms of teaching or curricula (Bello-Bravo et al., 2018b, 2018c). The medium's effectiveness includes high rates of solution adaptation two years after the information's initial presentation (Bello-Bravo et al., 2020a), and no loss of efficacy even when recipients culturally adapt (modify) the protocol in local ways (Bello-Bravo et al., 2020b).

The strength of this approach also lies in its media-specific affordances. Solutions offered embody the pragmatic, skills-centric, self-interested learning characteristic of effective adult education (Knowles, 1990), while the animated format itself readily supports *audio-visual multimodality*, facilitating information's simultaneous presentation across visual and auditory channels, reinforcing retention through redundancy (Bateson et al., 1956; Mayer, 2002; Nardone & Portelli, 2007). Such redundancy in communication, as Shannon and Weaver (1949) observed, reduces message senders' uncertainty about message reception—especially when feedback is not available (as when mass-scaling dissemination). By controlling message delivery on the producer side, and by designing for redundancy, animated content can reduce informational ambiguity more effectively than ad hoc instruction approach to communication (Bateson et al., 1956).

Redundancy also occurs through facilitated discussions following video exposure (Bello-Bravo et al., 2018c) and the ability of learners to control the pace of messaging, pausing or re-watching as needed, even in busy and distracted settings (Bello-Bravo et al., 2019). Conversely, in-person learning situations can both help and hinder knowledge transfer. In Bangladesh, women preferred women-led agricultural extension presentations (Medendorp et al., 2022), and Tata and McNamara (2016) found that women (in contrast to men) preferred face-to-face rather than computer-mediated instruction. A speaker's accent can affect a listener's perception of their intelligence or authority (Freynet & Clément, 2019), while stylized media—such as animation or puppetry—can introduce an aesthetic distancing that improves openness to content (Zics, 2011), enabling people to engage with difficult or politically sensitive material more openly. In some cases, animated presentations may be perceived as more neutral or trustworthy than live instructors, depending on context.

Lastly, as catalysts for lowering transactional costs, short, locally translated, generically animated educational videos provide an "edge" between the provider and recipient that increases the likelihood of adoption, especially around the basic SDGs of human dignity and food, health, shelter, and gender security. This is not to suggest that animation should be the sole channel for instruction. Like language, media must be culturally and contextually appropriate. A genuine commitment to universal empowerment will recognize a plurality of methods that work, and when one route becomes blocked, another must be sought. Baseline, however, is that a given human right must be achievable to be real. We would not accept that education, health, or safety were rights if we believed they were inherently unreachable. The same holds for the right of knowledge.

Designing Educational Content for Diverse Contexts

What is *gold-standard, life-sustaining knowledge*? *Gold-standard* knowledge means the best-available, empirically validated knowledge and practices—whether a food security technique to hermetically store postharvest beans, a technique for preventing the spread of an invasive insect or disease (COVID-19, tuberculosis), or strategically arranging stones and grates to reduce the use of firewood during cooking by 30–50%, thereby reducing forest degradation, CO_2 emissions, and the effects of climate change. What is particularly important for gold-standard knowledge is an ability to update it in a highly timely manner as improved information arises, whether by people doing basic research in formal settings or through communities of practice sharing experiences from the field.

The reference to empirically validated knowledge and practices is not in any way intended to denigrate traditional knowledge and practices; empiricism is by no means limited only to the scientific method (Agrawal, 1995; Bello-Bravo et al., 2019; Horwitz et al., 2017). Moreover, in some cases, traditional medicine has prompted the scientific community to examine and understand mechanisms in nature that unlock healing (Che et al., 2022; Jansen et al., 2021; Qadir & Raja, 2021). Curiosity will often want to understand *why* something works, and in no way would we suggest that someone should refrain from such research, above all because it might well disclose a further way to improve on what already works. We also recognize that science, like governance, can sometimes be very slow to act—even when facts on the ground (and in the air, like climate change) do not afford the luxury of years to arrive at certainty that our protocols and knowledge are as exact as they could be. Where that bar of certainty should be set is not itself always clear. For example, among the 89% of adapters of the jerrycan technique, one in three farmers skipped one of the eight steps of the recommended protocol, while one in five added a step (by including traditional additives to the stored beans) (Bello-Bravo et al., 2020b). Despite these "mistakes" in the protocol's application, none of the farmers reported a failure of the technique or that their stored beans had been ruined by bruchids.

Global ICT systems that aim to build a universal knowledge base offer more than dissemination; they also open a pathway for integrating ideas and observations that do not originate in academic laboratories. This should not be feared but welcomed. Drawing the global population into the empirical process of hypothesis testing, even at informal or experiential levels, can only enrich human knowledge across domains. Nor is this point to claim that traditional practices or local discoveries are automatically equivalent to rigorously tested scientific results. It is simply to observe that progress often begins with ordinary human observation. Solutions to persistent problems frequently arise from insights noticed in daily life, then refined through systematic investigation. For example, the discovery of beneficial microbes in agriculture began not in a lab, but in farmers' recognition that composting improved plant health and yield. Formal science followed—not replaced—that intuitive, experience-based insight (Hopwood, 2007).

There is broad consensus among creativity researchers that creativity is always a function of the knowledge base. The wider and more diverse that base, the more raw material innovators can draw on to construct new paradigms of understanding (Gerver et al., 2023). This same principle underlies the architecture of artificial intelligence, which builds novel outputs from vast and varied datasets (Grilli & Pedota, 2024). For this reason, one might argue artificial intelligence will soon surpass human innovation in certain spaces, not because its reasoning or creativity skills are better than humans, but because it is able to draw on a much broader knowledge base (Johannessen, 2023).

In this sense, the right to knowledge is not a just a moral obligation, it is to the benefit of all of humanity to unleash the creative powers of the entire human race by broadening the knowledge base from which it draws. Why would we purposely exclude entire segments of the population from the curiosity enterprise that we call science? The most amazing discoveries often come from the least likely sources. All of this reinforces the need to shift to an ICT-based solution system. By (finally) making the leap from in-person extension systems to cloud-based systems that have the potential to deliver knowledge anywhere at any time, we will be able to exponentially increase the innovations from the ground up.

The Multi-Country Impact of the SAWBO RAPID Project

As an example of the power of loosely coupled ICT platforms, in 2020 and 2021, through a collaboration with and support from USAID, the SAWBO *RAPID* program translated a jerrycan video into 100 additional language variants and distributed it through select media channels. These avenues included traditional media (television and radio), new media (social media, social networks, webpages, and list-servs), and traditional in-person systems, like the national agricultural extension system and NGOs. Regions in Ghana, Kenya, and Nigeria designated by USAID as "Zones of Influence" (ZOIs) were selected, as ZOIs are typically underdeveloped economic zones where USAID would focus its efforts and resources.

SAWBO *RAPID* reached a documented 12.5 million people in the three countries, based on Google analytics and triangulating with an estimate based on an on-the-ground survey using a randomized sampling of adults in the target areas (Fernández-Mesa et al., Under Review). The randomized sampling of adult viewers also found that, on average, 93.2% of viewers reported understanding the technique as explained in the video, 90.3% reported being confident that they could teach the technique to someone else, and 58.1% reported using the technique in their homes, farms, or businesses. These are unusually high numbers, especially for a "cold" share of video information without a dissemination agent to mediate the reception.

However, it must be stressed that this was not simply a matter of "putting a video on blast." Rather, it involved carefully curated, iteratively identified pathways for transmitting the video (including establishing new partnerships when planned channels proved unusable, requiring the identification of alternative pathways). Also, the video content had already demonstrated a high adoption rate, largely because it depicted a simple and actionable remedy for a problem identified by farmers themselves as critical, namely postharvest lost. The project was undertaken during COVID-19, when people were confined to their homes and farms and may have more actively been seeking information remotely as threats to food security were felt more urgently (due to mobility limitations and impacts to crop yields). Regardless of the reason for these numbers, what it did definitively show was that it was possible to reach a mass audience without on-the-ground personnel to mediate the information. It also shows that rates of use can be comparable to those of in-person training.

The objection that there is no guarantee that the technique is being applied properly also holds for in-person training after the team leaves. Follow-ups on in-person training rarely occur. ICT-based dissemination offers the advantage that if a refresher is needed or if one wishes to share the technique with others, the content is always available. The memory problem is a major barrier for in-person training (Murre & Dros, 2015). Even under ideal circumstances, most of what is received in in-person training has been lost to the memory within 10 days of training (see Fig. 4.3). The use of ICT-based content means that the accuracy of the technique does not rely on the retention of information conveyed by an expert but is available at any time for users to return, review, refresh, and reapply so that the accuracy of application is retained over time.

Consistent with the insight that distribution costs for ICT mass-scaling are radically lower than traditional in-person training, the per-person cost for SAWBO *RAPID*'s mass-scaling ranged from less than $0.10 per person to less than $1.80 per person for video full views in various languages (Fig. 4.4). We have since used AI tools to drop this cost further. On all dimensions—i.e., cost, reach, retention, and sharing—ICT-based solutions will always outperform in-person training. If we are to talk seriously of a right to knowledge, these ICT-based solutions are really the only option for attaining the goal of a universal knowledge base.

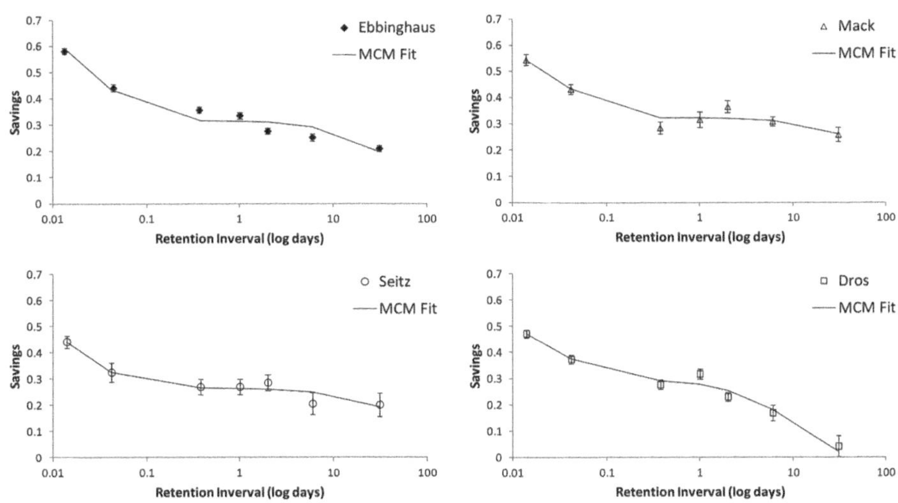

Fig. 4.3 Forgetting curves across four studies. This figure depicts four savings-based forgetting curves from different studies, plotted as a function of the retention interval (in log days). "Savings" refers to the proportion of relearning time saved compared to initial learning, providing a metric of memory retention. As the retention interval increases, savings decrease, illustrating the general pattern of forgetting over time. Reprinted with permission from Murre and Dros (2015)

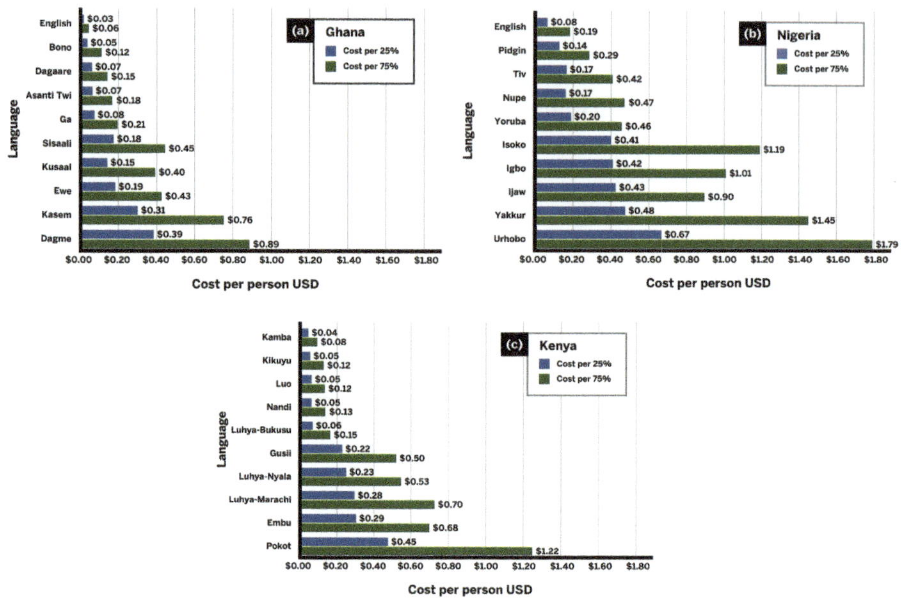

Fig. 4.4 Per person costs of YouTube video dissemination (advertisement pushes) during SAWBO *RAPID*. Costs are depicted for 25% and 75% viewing time of the jerrycan storage video across a diversity of languages. The costs are given by country: (**a**) Ghana, (**b**) Nigeria, and (**c**) Kenya. Since the SAWBO *RAPID* program, SAWBO has worked with collaborators to use AI tools to drive these costs down further. Figure reprinted with permission from Bello-Bravo et al. (2022)

ICT-Scaling as a Research Construct

Given the present-day ubiquity of digital ICTs and the reflexive assumption that they will comprise some part of *any* solution in every sector of life, it is startling that a Google Scholar search for research literature on "ICT-scaling" itself returns only 12 results; for example, Goyal et al. (2022) emphasize the ICT-scaling's cost-effectiveness. More research focuses on integrating digital ICTs into global systems, especially with the explosion of technologies for machine learning, artificial intelligence, simulated neural networks, large language model processing, and improved robotics.

However, ICT-scaling is a phenomenon subject to research in its own right. While it is always important to measure the penetration of ICTs into a particular domain, ICT-scaling more importantly analyzes how innovations in the form of information, knowledge, and practices that support life-sustaining results—such as the SDG benchmarks (Rodríguez-Domenech et al., 2023)—are effectively scaled-up to reach, engage, and empower wider audiences. While documenting decreasing unit costs to upscale is an important first step (Bello-Bravo et al., 2022), the methodological and quantitative findings from SAWBO *RAPID* also furnish key insights into how to establish partnerships for identifying distribution channels and measuring the engagement and use of the animated content. The issue at stake here is centrally people's exercise of the right to knowledge in order to "participate in an informed way in decisions that affect them" (UNESCO, 2010, p. 16).

Conclusion

At this historical juncture—when nearly 75 years of development practice under the Bretton Woods framework faces collapse with the loss of three out of every five aid dollars—we must fundamentally reassess how we approach knowledge sharing. ICT-based scaling systems clearly outperform traditional models in cost, reach, and retention, relying on pull factors rather than push. Their ability to increase scale while reducing cost makes it possible, for the first time, to genuinely aspire to a universal knowledge base capable of catalyzing global innovation.

Recent crises have only reinforced this urgency. COVID-19 disrupted supply chains and market access; the war in Ukraine has restricted agricultural inputs and outputs. As a result, progress toward the SDGs has stalled—and will remain out of reach unless we finally make the shift to ICT-based knowledge systems. This is not a rhetorical flourish but a demonstrated fact. There are no longer any legitimate excuses for failing to adopt these platforms to deliver life-sustaining, gold-standard knowledge at scale.

However, to make this shift effective, we must also develop empirical ways to assess engagement and empowerment. Without such measures, we cannot tailor strategies or ensure universal access to knowledge as a true human right. Still, among all options currently available, this trajectory is the only viable direction for transforming elite knowledge into widespread behavioral change. ICT-based systems must become the standard for all future extension work.

References

Adavant, S. (2021). *Now's the time to dump WhatsApp, privacy advocates say*. https://www.digitaltrends.com/mobile/is-now-the-time-to-dump-whatsapp/

AGRA. (2020). *Extension strategy: AGRA'S private sector-led approach to extension*. AGRA. https://agra.org/wp-content/uploads/2017/09/Final-AASR-2017-Aug-28.pdf

Agrawal, A. (1995). Dismantling the divide between indigenous and scientific knowledge. *Development and Change, 26*(3), 413–439. https://doi.org/10.1111/j.1467-7660.1995.tb00560.x

Ahad, A. D., & Lim, S. M. A. (2014). Convenience or nuisance?: The 'WhatsApp' dilemma. *Procedia-Social and Behavioral Sciences, 155*, 189–196. https://doi.org/10.1016/j.sbspro.2014.10.278

Attuh, S., & Kankam, P. K. (2024). Community radio as information dissemination tool for sustainable rural development in Ghana. *Journal of Radio & Audio Media, 31*(1), 248–270. https://doi.org/10.1080/19376529.2022.2146119

Bateson, G., Jackson, D. D., Haley, J., & Weakland, J. (1956). Toward a theory of schizophrenia. *Behavioral Science, 1*(4), 251–264. https://doi.org/10.1002/bs.3830010402

Bello-Bravo, J., & Pittendrigh, B. R. (2012). Scientific animations without Borders: A new approach to capture, preserve and share indigenous knowledge. *The Journal of World Universities Forum, 5*(2), 11–20. https://doi.org/10.18848/1835-2030/CGP/v05i02/56788

Bello-Bravo, J., Huesing, J., Boddupalli, P. M., Goergen, G., Eddy, R., Tamò, M., & Pittendrigh, B. R. (2018a). IPM-based animation for fall armyworm: A multi-institutional and virtual international collaboration using the Scientific Animations Without Borders (SAWBO) platform. *Outlooks on Pest Management, 29*(5), 225–230. https://doi.org/10.1564/v29_oct_10

Bello-Bravo, J., Tamò, M., Dannon, E. A., & Pittendrigh, B. R. (2018b). An assessment of learning gains from educational animated videos versus traditional extension presentations among farmers in Benin. *Information Technology for Development, 24*(2), 224–244. https://doi.org/10.1080/02681102.2017.1298077

Bello-Bravo, J., Zakari, O. A., Baoua, I., & Pittendrigh, B. R. (2018c). Facilitated discussions increase learning gains from dialectically localized animated educational videos in Niger. *Information Technology for Development, 24*(2), 1–25. https://doi.org/10.1080/02681102.2018.1485004

Bello-Bravo, J., Lutomia, A. N., Njoroge, T. M., & Pittendrigh, B. R. (2019). Beyond binaries of scientific and indigenous knowledge bean storage techniques: A case of market women in Ghana. In S. Hameed, S. el-Kafafi, & R. Waretini-Karena (Eds.), *Handbook of research on indigenous knowledge and bi-culturalism in a global context* (pp. 38–60). IGI.

Bello-Bravo, J., Abbott, E., Mocumbe, S., Mazur, R., Maria, R., & Pittendrigh, B. R. (2020a). An 89% solution adoption rate at a two-year follow-up: Evaluating the effectiveness of an animated agricultural video approach. *Information Technology for Development, 26*(3), 577–590. https://doi.org/10.1080/02681102.2019.1697632

Bello-Bravo, J., Abbott, E. A., Mocumbe, S., & Pittendrigh, B. R. (2020b). Identifying and evaluating farmer deviations from steps recommended for hermetic postharvest storage of beans in northern Mozambique. *Journal of Stored Products Research, 87*, 101628. https://doi.org/10.1016/j.jspr.2020.101628

Bello-Bravo, J., Brooks, I., Lutomia, A. N., Bohonos, J. W., Medendorp, J. W., & Pittendrigh, B. R. (2021). Breaking out: The turning point in learning using mobile technology. *Heliyon, 7*(3), e06595. https://doi.org/10.1016/j.heliyon.2021.e06595

Bello-Bravo, J., Medendorp, J. W., Lutomia, A. N., Reeves, N. P., Tamò, M., & Pittendrigh, B. R. (2022). Dramatically increased accessibility and decreased cost-per-person impacts are needed for scaling IPM in Africa. *Current Opinion in Insect Science, 54*, 100971. https://doi.org/10.1016/j.cois.2022.100971

Bello-Bravo, J., Medendorp, J. W., Lutomia, A. N., & Pittendrigh, B. R. (2023). *Gender, digitalization, and resilience in international development: Failing forward*. Routledge. https://doi.org/10.4324/9781003226383

Bernard, T., Makhija, S., Spielman, D. J., & Abate, G. T. (2019). *The (marginal) cost of technology adoption: A cost-effectiveness analysis of digital green's video-mediated agricultural extension approach in Ethiopia*. IFPRI.

Box, G. E., & Draper, N. R. (1987). *Empirical model-building and response surfaces*. John Wiley & Sons.

Brundtland, G., Khalid, M., Agnelli, S., Al-Athel, S., Chidzero, B., Fadika, L., Hauff, V., Lang, I., Shijun, M., & de Botero, M. M. (1987). *Report of the World Commission on Environment and Development: Our common future* (O.U. Press, Ed.). World Commission on Environment and Development.

Che, Q., Luo, T., Shi, J., He, Y., & Xu, D.-L. (2022). Mechanisms by which traditional Chinese medicines influence the intestinal flora and intestinal barrier. *Frontiers in Cellular and Infection Microbiology, 12*, 863779. https://doi.org/10.3389/fcimb.2022.863779

Chen, J. J. (2010). The Hardy-Weinberg principle and its applications in modern population genetics. *Frontiers in Biology, 5*, 348–353. https://doi.org/10.1007/s11515-010-0580-x

Corrigan, P. W. (2020). Community-based participatory research (CBPR), stigma, and health. *Stigma and Health, 5*(2), 123–124. https://doi.org/10.1037/sah0000175

Davis, K., Franzel, S., & Spielman, D. J. (2019). *Extension options for better livelihoods and poverty reduction: A selected review 2012–2015* (Vol. 3). Gates Open Research.

De Framond, A. J., Barton, K. A., & Chilton, M.-D. (1983). Mini–Ti: A new vector strategy for plant genetic engineering. *Bio/Technology, 1*(3), 262–269. https://doi.org/10.1038/nbt0583-262

Feder, A. (2024). *Countries by estimated nominal GDP in 2025*. https://en.wikipedia.org/wiki/List_of_countries_by_GDP_%28nominal%29#/media/File:Countries_by_GDP_(nominal).svg

Fernández-Mesa, V., Lutomia, A. N., Adekoya, A., Medendorp, J. W., Bello-Bravo, J. & Pittendrigh, B. R. (Under Review). *Scaling across media platforms: An animated video goes a long way.* Evidence from Nigeria.

Freynet, N., & Clément, R. (2019). Perceived accent discrimination: Psychosocial consequences and perceived legitimacy. *Journal of Language and Social Psychology, 38*(4), 496–513. https://doi.org/10.1177/0261927X19865775

Gerver, C. R., Griffin, J. W., Dennis, N. A., & Beaty, R. E. (2023). Memory and creativity: A meta-analytic examination of the relationship between memory systems and creative cognition. *Psychonomic Bulletin & Review, 30*(6), 2116–2154. https://doi.org/10.3758/s13423-023-02303-4

Goyal, D., Soni, Y., & Gandhi, G. (2022). A study on e-health infrastructure in rural India. *ECS Transactions, 107*(1), 19299. https://doi.org/10.1149/10701.19299ecst

Grilli, L., & Pedota, M. (2024). Creativity and artificial intelligence: A multilevel perspective. *Creativity and Innovation Management, 33*(2), 234–247. https://doi.org/10.1111/caim.12580

Gumucio, T., Hansen, J., Huyer, S., & Van Huysen, T. (2019). Gender-responsive rural climate services: A review of the literature. *Climate and Development, 12*(3), 241–254. https://doi.org/10.1080/17565529.2019.1613216

Hopwood, D. A. (2007). *Streptomyces in nature and medicine: The antibiotic makers*. Oxford University Press.

Horwitz, R. I., Hayes-Conroy, A., Caricchio, R., & Singer, B. H. (2017). From evidence based medicine to medicine-based evidence. *The American Journal of Medicine*. https://doi.org/10.1016/j.amjmed.2017.06.012

Huyer, S. (2016). Closing the gender gap in agriculture. *Gender, Technology, and Development, 20*(2), 105–116. https://doi.org/10.1177/0971852416643872

Iacoella, F., Gassmann, F., & Tirivayi, N. (2022). Which communication technology is effective for promoting reproductive health? Television, radio, and mobile phones in sub-Saharan Africa. *PLoS One, 17*(8), e0272501. https://doi.org/10.1371/journal.pone.0272501

Jansen, C., Baker, J., Kodaira, E., Ang, L., Bacani, A., Aldan, J., Shimoda, L., Salameh, M., Small-Howard, A., & Stokes, A. (2021). Medicine in motion: Opportunities, challenges and data analytics-based solutions for traditional medicine integration into western medical practice. *Journal of Ethnopharmacology, 267*, 113477. https://doi.org/10.1016/j.jep.2020.113477

Johannessen, J.-A. (2023). *Consciousness and creativity in artificial intelligence: The cognitive side of knowledge management*. Emerald Publishing Limited.

Kabeer, N. (1999). Resources, agency, achievements: Reflections on the measurement of women's empowerment. *Development and Change, 30*(3), 435–464. https://doi.org/10.1111/1467-7660.00125

Khatri, A., Lallawmkimi, M. C., Rana, P., Panigrahi, C. K., Minj, A., Koushal, S., & Ali, M. U. (2024). Integration of ICT in agricultural extension services: A review. *Journal of Experimental Agriculture International, 46*(12), 394–410. https://doi.org/10.9734/jeai/2024/v46i123146

Knowles, M. S. (1990). *Andragogy in action: Applying modern principles of adult learning*. Jossey-Bass Publishers.

Komodromos, M. (2024). Gender barriers experienced by women entrepreneurs in Cyprus. In M. Chhabra (Ed.), *Women entrepreneurship policy: Context, theory, and practice* (pp. 169–186). Springer.

Lowe, R. K. (2001). *Beyond "eye-candy": Improving learning with Animations Apple University Consortium Academic and Developers Conference*, Townsville, Queensland, Australia.

Lowe, R. K. (2003). Animation and learning: Selective processing of information in dynamic graphics. *Learning and Instruction, 13*(2), 157–176. https://doi.org/10.1016/S0959-4752(02)00018-X

Lutomia, A. N., Bello-Bravo, J., Lutomia, N. I., Kataru, J. K., Medendorp, J. W., & Pittendrigh, B. R. (2024). Participatory practices and lessons from Scientific Animations Without Borders and a WhatsApp network in a post-COVID age: The case of video animations for rural agriculture. In A. Chowdhury & G. A. Gow (Eds.), *Digital communication for agricultural and rural development* (pp. 253–270). Routledge.

Mayer, R. E. (2002). Multimedia learning. *Psychology of Learning and Motivation, 41*, 85–139. https://doi.org/10.1016/S0079-7421(02)80005-6

Medendorp, J. W., Reeves, N. P., Sal y Rosas Celi, V. G., Harun-ar-Rashid, M., Krupnik, T. J., Lutomia, A. N., Pittendrigh, B., & Bello-Bravo, J. (2022). Large-scale rollout of information and communication technology-enhanced extension training in Bangladesh demonstrates challenges and opportunities towards inclusive gender participation. *PLoS One, 17*(7), e0270662. https://doi.org/10.1371/journal.pone.0270662

Mocumbe, S. (2016). *Use of animated videos through mobile phones to enhance agricultural knowledge among bean farmers in Gurúè District, Mozambique* [Masters Thesis, Iowa State University]. Ames, IA.

Murre, J. M., & Dros, J. (2015). Replication and analysis of Ebbinghaus' forgetting curve. *PLoS One, 10*(7), e0120644. https://doi.org/10.1371/journal.pone.0120644

Nardone, G., & Portelli, C. (2007). Caught in the middle of a double-bind: The application of non-ordinary logic to therapy. *Kybernetes, 36*(7/8), 926–931. https://doi.org/10.1108/03684920710777432

Nchake, M. A., & Shuaibu, M. (2022). Investment in ICT infrastructure and inclusive growth in Africa. *Scientific African, 17*, e01293. https://doi.org/10.1016/j.sciaf.2022.e01293

Omanga, D. (2019). WhatsApp as 'digital publics': The Nakuru Analysts and the evolution of participation in county governance in Kenya. *Journal of Eastern African Studies, 13*(1), 175–191. https://doi.org/10.1080/17531055.2018.1548211

Ortega, M., & Lin, L. (2004). Control theory applications to the production–inventory problem: A review. *International Journal of Production Research, 42*(11), 2303–2322. https://doi.org/10.1080/00207540410001666260

Payumo, J., Bello-Bravo, J., & Pittendrigh, B. R. (2021). Demonstrating the nexus effects of online videos, research outputs, and investments to knowledge absorption using linguistically adapted animations. *European Journal of Applied Sciences, 9*(2), 283–301. https://doi.org/10.14738/aivp.92.10025

Qadir, S. U., & Raja, V. (2021). Herbal medicine: Old practice and modern perspectives. In R. A. Bhat, K. R. Hakeem, & M. A. Dervash (Eds.), *Phytomedicine* (pp. 149–180). Academic Press. https://doi.org/10.1016/B978-0-12-824109-7.00001-7

Rajasekhar, D., Babu, M. D., & Manjula, R. (2018). Elite capture in decentralised institutions: A literature survey. In D. Rajasekhar, M. D. Babu, & R. Manjula (Eds.), *Decentralised governance, development programmes and elite capture* (pp. 19–29). Springer. https://doi.org/10.1007/978-981-13-1900-6_2

Reeves, N. P., Sal y Rosas Celi, V. G., Lutomia, A. N., Medendorp, J. W., Bello-Bravo, J., & Pittendrigh, B. (2024). Agricultural education in Africa using YouTube multilingual animations: A retrospective feasibility study assessing costs to reach language-diverse populations. *PLoS One, 19*(4), e0302136. https://doi.org/10.1371/journal.pone.0302136

Rodríguez-Domenech, M. Á., Bello-Bravo, J., Lutomia, A. N., Medendorp, J. W., & Pittendrigh, B. R. (2023). Digital education and community outreach for sustainable development in Africa. In W. Leal Filho, I. R. Abubakar, I. da Silva, R. Pretorius, & K. Tarabieh (Eds.), *SDGs in Africa and the Middle East Region: Implementing the UN sustainable development goals regional perspectives* (pp. 1–31). Springer. https://doi.org/10.1007/978-3-030-91260-4_65-1

Shannon, C. E., & Weaver, W. (1949). *The mathematical model of communication*. University of Illinois Press.

Smith, D., McLaughlin, T., & Brown, I. (2012). 3-D computer animation vs. live-action video: Differences in viewers' response to instructional vignettes. *Contemporary Issues in Technology and Teacher Education, 12*(1), 41–54.

Tata, J. S., & McNamara, P. E. (2016). Social factors that influence use of ICT in agricultural extension in southern Africa. *Agriculture, 6*(2), 15. https://doi.org/10.3390/agriculture6020015

Tata, J. S., & McNamara, P. E. (2018). Impact of ICT on agricultural extension services delivery: Evidence from the Catholic relief services SMART skills and Farmbook project in Kenya. *The Journal of Agricultural Education and Extension, 24*(1), 89–110. https://doi.org/10.1080/1389224X.2017.1387160

UNESCO. (2010). *Freedom of information: The right to know*. UNESCO.

Williams, A., & Tkach, B. K. (2022). Access and dissemination of information and emerging media convergence in the Democratic Republic of Congo. *Information, Communication & Society, 25*(10), 1383–1399. https://doi.org/10.1080/1369118X.2020.1864004

Zics, B. (2011). Toward an affective aesthetics: Cognitive-driven interaction in the affective environment of the mind cupola. *Leonardo, 44*(1), 30–37. https://doi.org/10.1162/LEON_a_00090

Chapter 5
How ICT-Based Big Data, Machine Learning, and AI Support the Achievement of the Sustainable Development Goals: The Case of SAWBO

Introduction

Three terms have recently entered the lexicon of international development, marking advances in data generation, analysis, and what is sometimes called cognitive computing: artificial intelligence (AI), big data, and machine learning. Like many trends in the field, these terms have quickly become buzzwords, often used to accessorize mediocre funding proposals or exoticize otherwise ordinary research results (Sahoo, 2021). While these tools offer genuine promise, they are frequently misunderstood and misapplied. This section briefly defines each before examining how they are, can be, and will be integrated into SAWBO's ongoing growth and expansion.

Big Data

The new era in information generation and knowledge creation started with the "datafication" of daily life: the translation of every part of our existence into machine readable information or big data (Cukier & Mayer-Schoenberg, 2013). The world's volume of data is increasing dramatically. In 2020, 64.2 zettabytes of data (or 64.2 trillion gigabytes) were created—a 314% increase from 2015. In 2024, the global volume of data was 149 zettabytes. By the end of 2025, the global volume of data is projected to rise further to 181 zettabytes (Bartley, 2024). Much is "data exhaust"—traces generated through interactions with digital systems, including passive, secondary data (e.g., metadata, clickstreams, location data, device usage patterns, and log files) that are not the primary purpose of the interaction but are captured anyway (O'Leary & Storey, 2022). This is a scientist's dream, or nightmare, since today's data generation can make yesterday's observations obsolete.

J. Bello-Bravo et al., *The Right to Knowledge*,
https://doi.org/10.1007/978-3-032-03360-4_5

With the rise of electronic computing, electronic networks, data storage capability, and improved data analysis tools came the ability to generate, organize, and quickly analyze massive amounts of data that would be inaccessible to the unaided human brain. In statistical analysis, this expansion presented new opportunities for the emerging field of data science and large-scale probabilistic hypothesizing that will allow us to work with increasing levels of confidence in understanding human and non-human activity and even make predictions based on the understanding (Ghahramani, 2015; Vershynin, 2018). However, van Es and Masson (2018) caution against calling big data a revolution, noting that it is really a change in our capacity to search, aggregate, and cross-reference large datasets: "a capacity that rests on the interplay of technology (computer power, algorithmic accuracy), analysis (the ability to draw patterns from data that are economically, socially, or technically significant) *and* mythology (the belief that large datasets offer a 'higher form of intelligence'" (p. 1).

Research on big data over the last two decades is usually viewed in one of the two ways. On the one hand, it is viewed as a data deluge, an unprecedented and unforeseen increase in the volume of information available for analysis (Nazeer & Nadir, 2024). This view emphasizes its chaotic and uncontrollable status but also suggests that it is rich in possibilities, leading one writer to declare "the end of theory" (Anderson, 2008, p. 16), opining that probabilistic computing could replace experimental science and theoretical hypothesis generation. This scenario has not happened, and given how the field has evolved since the high-water mark of enthusiasm or hype about big data in the late 2000s, we affirm the persistent importance of human-driven theory, experimentation, and the scientific method in driving true innovation and understanding.

Nonetheless, big data has definitely changed the way we view information. Experts involved in big science often accept that the emergence of big data has been unruly but that—through collaborative and well-planned data collection efforts driven by the international scientific community—order and rigor can be brought to the data deluge. Gains like those advancements in genomics and NEON were achieved (Zimmerman & Nardi, 2016). In this view, big data presents opportunities to expand our approaches to science and further emergence in the field of data science to help solve wicked problems like those targeted by the SDGs, through opening new fields of data intensive research (Leonelli, 2021) and data-driven decision-making (Sarker, 2021). Of course, both of these descriptors are correct. There is no controlling the number of data generation points now or in the future, nor should there be. However, as is currently taking place, there is still need for systematic distinctions between data types and methodologies so that analytical outcomes are derived through the rigorous application of sound scientific principles.

Machine Learning

Machine learning is a derivative of big data and a subset of AI. It refers to the continuous improvement of systems through the analysis of big data and the application, in some cases in real time, of the results of that analysis based on better information and improved knowledge. The functions of machine learning consist of detection, recognition, diagnosis, optimization, and prediction (Tang et al., 2021). It is not as though the principles of machine learning have not been present in systems theory before. Some ascribe the principles to the early deciphering work of Alan Turing in the Enigma Project during World War II. While that specific project can certainly be debated, the resulting Turing Test for artificial intelligence provided a basis for machine learning science already by the 1950s (Alzubi et al., 2018).

What is new in machine learning is the scale and speed with which data can be processed and applied to systems so that improvements can be programed into the analysis itself and can make constructive changes to systems almost in real-time. One such example of machine learning is self-driving cars. Continuous improvement is programmed into the processing of the data generated by the system for ever safer versions of itself (Ni et al., 2022; Stilgoe, 2018; Wang et al., 2024). Continuous systems improvement constitutes a dramatic step forward for data analysis, with machine learning as the core process to drive it.

AI

There is no complete consensus around what AI is, but it generally involves algorithms that imitate patterns of human thought, speech, decision-making, and reasoning. While machine learning is a form of artificial intelligence—in that within the parameters of a specific system, the system can learn from and adapt to changes in its environment to optimize its function—it is not usually programmed to move beyond that system and incorporate possible learning from outside the system. AI, however, incorporates a number of other capabilities—including deep learning, natural language process, computer vision, and neural networks—that can surpass the confines of any specific system (the machine) and its function, to explore other systems and their potential influence on the functioning of one particular system.

To take the self-driving car again as an example, machine learning allows the car to compile data on internal and immediate environmental factors affecting the system to optimize its function, which is to drive safely and efficiently in traffic while delivering the passenger to the desired destination. In contrast, a fuller spectrum of AI capabilities would enable it to decipher a vast array of sources to consider not only driving safely and efficiently to a destination but also to find traffic patterns in its environment, identify faster routes, suggest other forms of transportation, or even design a better car. It could also, in theory, do research on cars' impacts on the environment, analyze social factors influencing driving behavior, predict the chances that you will sustain an accident on any given day, at any given time, or in any given place. While machine learning can optimize one system, AI could optimize meta-systems by balancing the optimization of several systems simultaneously for the "greater good."

In this sense, AI not only holds the potential to sense what is happening within a system in real-time but also to explore a vast number of complexities, past or present, that are not necessarily reducible to data points, thus emulating patterns in human experience that we recognize as resembling thinking, reasoning, hypothesizing, experimenting, and even fantasizing, dissimulating, and fabricating. We must be circumspect about over-reifying these resemblances.

Humans invent technologies that reflect their experiences of the world. For example, it took many experiments and trial and error to develop motion picture technologies (and their precursor cameras) that aligned with how the average human visual system perceives the world (Dyer, 1997). Early film technology, including lighting, film stock, and cinematography, was designed with white skin tones in mind. As a result, people of color, especially Black people, were depicted inaccurately, underexposed, or in ways that perpetuated racist stereotypes, and it would be decades before people of color appeared correctly in film (Dyer, 1997).

A technology that does not reflect our experience of the world will likely be discarded or superseded as a failure, just as early attempts at cameras were deemed failures. It is no wonder that AI is deemed legitimate if it appears to convincingly replicate human experience closely enough, as indicated in the fact that earlier attempts were abandoned as faulty. However, this replication is no more evidence of artificial thinking, reasoning, hypothesizing, experimenting, fantasizing, dissimulating, or fabricating than a camera taking a picture is an example of "seeing." Recognizing this dissimilarity does not erase the ethical implications that AI poses; rather, it refocuses attention on where the peril lies, and it is not that AIs will supersede or overthrow humanity. Rather, the risk crops up in how AI might take human governance decisions out of human hands or cause situations where decisions reached by AI algorithms cannot be replicated, reality-checked, or even tested.

Big Data, Machine Learning, AI, and the United Nations SDGs

AI, machine learning, and big data are inherently interconnected, with each enhancing the capabilities of the other. AI relies on big data as a foundational input to learn and refine decision-making processes. By leveraging vast datasets, particularly those groupings in machine learning, AI tools can analyze patterns and trends impossible for humans to discern, enabling more accurate predictions and insights. Big data serves as the essential fuel for AI, providing the raw material needed to train algorithms, which in turn organize and turn the data into actionable insights. As digital transformation accelerates, big data and AI integration is reshaping industries by revealing unexpected insights, changing decision processes, and opening new avenues for innovation.

Cognitive computing models, powered by machine learning, further enhance the ability to extract meaningful information from extensive datasets, surpassing conventional methods in efficiency and accuracy.

All of these technologies can be helpful for reaching the SDGs. For example, in agriculture, big data and AI integration can reshape capacity development for farmers, particularly in the Global South (Jamba & Marambi, 2024). AI improves knowledge sharing, facilitates transformative learning through dialogue and critical reflection, and automates repetitive tasks, freeing up time for more impactful activities. It also supports formation of virtual communities of practice (Cambridge et al., 2024), enabling farmers to collaborate, share insights, and adopt innovative techniques. By leveraging these technologies, farmers can develop skills to adapt to evolving agricultural challenges, improve food security, and participate in technological advancements. Ultimately, the synergy between AI and big data empowers individuals and organizations to achieve high-quality outcomes and meet the demands of the information age.

The real challenge here is defining the "greater good." It is important to keep in mind that AI, like any technology, reflects the assumptions of its creators. Programmers write the algorithms with which AI functions, and AI will always reflect not only their personal values and priorities but their cultural ones (Fountain, 2022), even as they cannot retrace or understand how it arrives at its outputs, i.e., the black box issue (Brożek et al., 2024; Xu & Shuttleworth, 2024). This reflection partly explains how one can use different AI programs (say, ChatGPT and Claude) and receive significantly different answers to the same prompt (Eghosa, 2025), and how, as Tian et al. (2021) indicate, "Multidimensional face representation in a deep convolutional neural network reveals the mechanism underlying AI racism" (p. 1). Algorithms tell AI where to seek information, how to prioritize it, how to assess its validity, how to rank information in terms of importance, value, and accuracy, and, above all, how to present the outcome of those factors processed. Nonetheless, AI's programming also reflects the biases of its programmer (Fountain, 2022; Noble, 2018).

This unfairness obviously can be problematic. If AI is to be used for international development, we would first need to ensure that AI is programmed to reflect the values of the international development community. The difficulty here is that no consensus exists around methods and outcomes. We may be united in terms of what the SDGs should be, but there are different visions for the ideal conditions to which we aspire. For example, let us take the different visions of the economic growth school versus the environmental sustainability school. Whereas they might agree on the outcomes expressed in the SDGs, their approaches widely diverge. Any effort to reduce international development to AI will ultimately result in multiple AI tools programmed differently and reflecting the values of their programmers. And maybe this variation is not bad, as it could permit differing approaches to be validated by outcomes.

Acquiring the ability to analyze large datasets, install systems that learn from data to make improvements in global socioeconomic systems, or even to review human experience and think creatively about other approaches and solutions to optimize system functions in delivering on its desired goals holds great potential for international development and the achievement of the SDGs. The United Nations (UN) and other international organizations fill a pivotal role in establishing principles and standards to guide the ethical and effective use of AI tools humanitarian development purposes. These tools aim to enhance data utility through greater openness and transparency, safeguard privacy, and prevent human rights abuses. They also strive to mitigate inequalities in the production, access, and application of data, information, and knowledge. Achieving the SDGs in the digital era necessitates not only preventing the misuse of these tools but also their responsible use for the public good.

The Secretary-General's Independent Expert Advisory Group on a Data Revolution for Sustainable Development (IEAG) has recommended a UN-led initiative to mobilize the data revolution (IEAG, 2014). Key actions include fostering innovation to address data gaps, mobilizing resources to address global inequalities, and improving leadership and coordination to ensure data plays a transformative role in sustainable development. Increasingly, UN agencies are adopting big data analytics for operational and humanitarian purposes, guided by the UN Development Group's broad principles on data privacy, protection, and ethics. Since then, there have been several proposals regarding the potential use of AI tools for achieving the SDGs (Bachmann et al., 2022; Nedungadi et al., 2024; Omodei et al., 2023; Teh & Rana, 2023).

The UN World Data Forum, first convened in 2017, has become a critical platform for exploring how data can support sustainable development. These forums, which include diverse stakeholders from governments, civil society, academia, and the private sector, have produced outcome documents that track progress and set ambitions for the global data community. For instance, the Cape Town Global Action Plan (2017) and subsequent declarations, including the Dubai Declaration (2018), Bern Data Compact (2021), and Hangzhou Declaration (2023), emphasize funding mechanisms, the role of official statistics, and the recommitment to advancing sustainable development data systems. Further, in response to challenges such as the COVID-19 pandemic, these initiatives have highlighted the critical role of National Statistical Offices (NSOs) and the need for a robust data ecosystem to address global crises. Through a mix of in-person, virtual, and hybrid forums, the UN has expanded its global data community to over 20,000 stakeholders, demonstrating its commitment to inclusivity and innovation.

Global Pulse, an innovation initiative under the UN Secretary-General, focuses on leveraging data science for sustainable development and humanitarian efforts. Operating Pulse Labs in Indonesia, Uganda, and the United States, Global Pulse collaborates with UN and government partners to accomplish impactful analytics solutions. The initiative aims to lower barriers to adopting and scaling data innovations by promoting awareness and developing privacy-protective tools. To expand access to insights from big data across various industries, Global Pulse has pioneered collaborations with the private sector to operationalize "data philanthropy."

This approach involves companies sharing their data safely and responsibly for sustainable development and humanitarian purposes. A notable example is Global Pulse's 2016 partnership with the social media platform Twitter (now X). With hundreds of millions of posts sent daily in multiple languages, social conversations provide real-time information on diverse issues, such as food prices, job availability, healthcare access, education quality, and environmental disaster reports. This collaboration enables UN development and humanitarian agencies to transform publicly available data into actionable insights to support communities worldwide.

Additional partnerships demonstrate the potential of big data in driving positive change. For instance, the GSMA's "Big Data for Social Good" initiative utilizes the data capabilities of mobile operators to address humanitarian crises, including epidemics and environmental disasters. Similarly, the "Data for Climate Action" competition connected researchers globally with data and tools from leading companies to develop innovative, data-driven climate solutions. Another example involves the emerging concept of data collaboratives, a model that extends beyond conventional public–private partnerships (Ruijer, 2021). These collaborations involve stakeholders from various sectors, particularly companies, exchanging data to produce public value. Together, these initiatives illustrate how big data partnerships can advance sustainable development and address pressing global challenges effectively. Ultimately, the challenge is not simply to adopt new technologies but to embed them within a framework that recognizes knowledge as a human right—not a commodity, not a byproduct of economic systems, and certainly not the proprietary output of opaque algorithms. AI, big data, and machine learning will not deliver the SDGs unless they are harnessed by inclusive, accountable institutions and informed by the lived realities of those they are meant to serve.

That means confronting the digital divides not as a technical glitch but as a symptom of deeper structural inequalities—economic, educational, political, and epistemic. The goal must be not just to use AI to reach development targets but to ensure that people have the power, skills, and access to shape the tools being used on their behalf. Otherwise, we risk reproducing the very injustices these technologies promise to resolve. The question is no longer whether AI can help achieve the SDGs. The question is: *who gets to decide how it's used, and for whose benefit?* The promise and pitfalls of these partnerships are more than theoretical. The following section describes how SAWBO has actively engaged machine learning and other AI tools for global development impact.

SAWBO Case Studies Utilizing Machine Learning and AI tools

Bangladesh

One example (already partly discussed in "A Time, a Place, and a Means" in Chap. 3) used machine learning to explore a large dataset of 1070 "pop-up" agricultural extension training sessions. Hosted by the Agricultural Advisory Society (AAS) in partnership with the International Center for the Improvement of Wheat and Maize (CIMMYT) in four of Bangladesh's eight administrative regions, these pop-ups reached ~131,000 farmers (averaging 123 participants per presentation). SAWBO was unaware of this project in advance, but one of its videos was utilized as training material for the sessions' integrated pest management practices to prevent the spread of fall armyworm (SAWBO, 2018), a serious threat to agriculture in Bangladesh. SAWBO also did not take part in the data collection and only learned of the dataset afterward. Nonetheless, its dataset proved helpful for analyzing the effects of variables like the trainer's gender, time of day, day of the week, month, type of venue, and geographic location on (1) the expected number of extension event attendees and (2) the odds of female attendance, conditioned on the total number of attendees.

In a first study by Medendorp et al. (2022), strong gender-specific training preferences were identified. Factors that increased overall participation—such as male-led events held after 3:30 p.m. in Rangpur—tended to reduce female attendance. In contrast, conditions that supported female participation, such as certain times of day and specific venue types, often resulted in smaller total audience sizes overall. These patterns were shaped by cultural norms: women expressed a reluctance to go out alone at night, even to familiar locations, and were least likely to attend events in markets—likely because limited time for shopping made it impractical to pause for video viewings. They were most likely to participate in pop-ups held at gender-separated religious institutions. Qualitative data reinforced these trends. One woman explained she would attend evening trainings if she didn't have to go alone to hire men for her shop the next day. A male participant observed that women in his administrative district were less likely to attend than those in neighboring districts, attributing the difference to stricter enforcement of gender roles.

This gendered variance poses a dilemma for extension agents: training events designed to increase women's participation at certain times and places might decrease overall attendance, while events designed to attract more participants overall tend to draw fewer women. Given limited resources to offer pop-ups, this dilemma seems to force a choice between mitigating gender inequalities (SDG 5) with fewer attendees overall or reaching a wider audience (SDG 2), risking the exacerbation of existing gender inequalities and not closing the digital gender divide.

A subsequent study (Reeves et al., 2023) applied machine learning to the earlier study's dataset to model whether "machine-supported decision-making could improve training attendance and gender inclusivity, which would then justify its use in designing large-scale agricultural training programs" (p. 7). It found that "if gender inclusivity goals are sensibly designed using machine-supported decision-making, total attendance and female participation *can be concurrently increased*" (p. 14, emphasis added), and that

> machine-supported decision-making appears to be a viable tool for improving agricultural training attendance and gender inclusivity. By increasing training turnout and gender inclusivity, this machine learning application is expected to enhance the social and economic health of developing countries and the health of their populations (p. 10).

However, such output would depend on the development of continuous data collection on training events to improve the optimization of the variables, leading to more female participation and more participation overall. These two studies, carried out in conjunction with the deployment of SAWBO's training videos, show the power of ICT-based extension for improving delivery of training materials. With ever greater understanding of user needs, machine learning offers the opportunity to make world-class scientific knowledge available to even underserved groups (see Fig. 5.1). No less significantly, this mechanism demonstrates that it is not always necessary to make unsustainable "trade-off" decisions (Gibson, 2006), e.g., whether to support SDG 5 or SDG 2. By deft modeling, one can find a sweet spot to support both at cost-per-delivery rates well within the possible, even in the new era of massively changing international aid funding.

Ghana, Nigeria, and Kenya

A second use of new AI tools involved the distribution of video content through social media platforms during the COVID-19 pandemic. As noted previously, the SAWBO *RAPID* program mass-distributed a postharvest loss prevention video

Fig. 5.1 Bridging gold-standard knowledge. SAWBO takes gold-standard, scientifically validated knowledge and empowers people to connect with it practically, making it conveniently available whenever and wherever they want to access it, in formats and languages they can use, virtually anywhere on the planet. Reprinted with permission from SAWBO (2025)

during the pandemic to nearly 12.5 million end-recipients by a variety of media and non-media channels. Among the strategies used to identify channels for mass-distributing, one included relying on social media algorithms (via Google ads, Facebook, etc.) to introduce the content into people's feeds. While this influx was a "passive" adoption of Google's and Facebook's big data analytics of social media, the approach demonstrated the feasibility of delivering digital content to mass audiences at remarkably low cost, in some cases for less than 10 cents/full-video view (Bello-Bravo et al., 2022).

Since then, SAWBO has been experimenting with refining placement of the videos in feeds, based on the profile of a probable user using machine learning tools. These profiles are built over time (and vary according to various video content given the information derived from Google Analytics to determine what people in what regions are most likely to view the entire video). AI algorithms can then be used to ascertain where future social media distribution campaigns can be strategically placed. Through this approach, SAWBO has been able to achieve placement of video content into the feeds of viewers most likely to take an interest in the topic could range from 3.7 cents/full view or 1 cent for a 30% view.

The SAWBO App

A major challenge posed by the digital divide involves the infrastructural barriers that block people's access due to limited or no Internet access, geographical remoteness, no electricity, or a lack of technology to access or view digital content. However, any ICT-based system aimed at promoting people's right to knowledge and achieving the SDGs requires strategies to transition online content to offline, regardless of the offline content. SAWBO's app is specifically designed to make that transition (see Fig. 5.2).

The app continuously updates its content and language variants, allowing devices with pre-loaded materials to be returned for new content. This system has shown success in places like the Democratic Republic of the Congo and Nepal, where communities have adopted new technologies alongside using the app.

The app also tracks usage data, such as video views, downloads, and shares, which helps developers assess its effectiveness (see Fig. 5.3). Machine learning or AI tools are applied to this data, allowing SAWBO to predict user needs and optimize content distribution, significantly reducing the cost of targeting the right audiences. This process respects user privacy, collecting only location data with explicit consent.

By combining large datasets and applying machine learning and AI, SAWBO continuously improves its delivery of content to key communities. With approximately 7000 languages spoken worldwide, this approach lays the foundation for advancing the right to knowledge and achieving the SDGs.

As for other widespread digital access device types in digitally saturated contexts, emerging AI and machine learning techniques and leveraging of big data

Fig. 5.2 Using the SAWBO mobile phone app. (1) People and groups can access the complete online SAWBO library using this mobile phone app (left). (2) They can then filter the country, language, and topic video list to select and download videos that interest them onto the app (middle). (3) Downloaded videos then are available for offline viewing and sharing, either by directly displaying them on video-enabled phones or via Bluetooth® transfer to other devices (right). Reprinted with permission from SAWBO (2025)

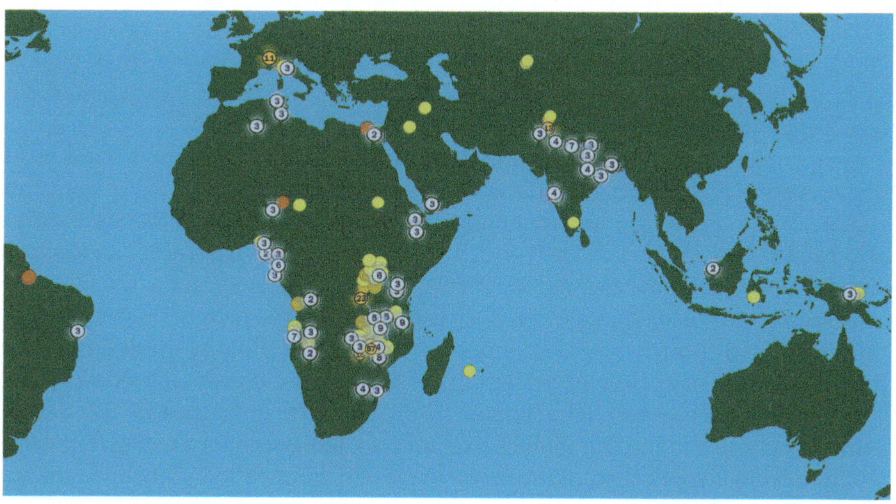

Fig. 5.3 Sample SAWBO app use during a 48-h period. This heat map depicts the geolocated activity of global mobile app users over a 48-h period in May 2025. Locations are shown only for users who explicitly consented to share their data; an additional 143 users (not shown) opted out of location sharing. Dot colors are varied for visual clarity only. Reprinted from SAWBO (2025)

already enable personalized content delivery pushes (potentially new content notifications on the SAWBO app), predictive analytics (to identify change agents and communities who could benefit from SAWBO content), language translation, and real-time feedback. These tools optimize user engagement by tailoring content to personal preferences and behaviors and can be further integrated into future iterations of the SAWBO app.

However, two major differences separate the SAWBO app from these increasingly widespread applications. First, while the app could draw on similar AI-driven strategies, rather than monetizing user behavior or pushing ads, the app leverages these technologies to provide a means for communities in remote or underserved areas access to personalized, context-aware educational content. Moreover, it does so without needing to fund its activities using an ad-based model because it already operates so cost-effectively. Further, by using AI to build profiles, to assist in distributing the app, and allowing change agents and communities with access to this platform, regardless of a person's gender, race, or geographic isolation, the SAWBO app enables critical available knowledge to be easily, conveniently, and usably pulled by these change agents where it is most needed—without the need for provider-side infrastructures to be funded and built or extension agents to be budgeted, hired, trained, and prepared to venture out into the field. More precisely, already underfunded extension agents will be much more empowered (with a single tablet, solar charger, and enough gasoline to travel from village to village) to deliver content, and where no extension agents exist, even lone change agents can take up that role.

One major caveat must be emphasized. A technical hurdle blocking an exponential reach and distribution of SAWBO content (especially in Africa) involves translating its videos into the most relevantly used dialect or language in an area. As such, the possibility of large language model (LLM) automated translation of content is truly the golden goose that could lay 7000 golden eggs for all of the world's language. However, inaccuracies and imprecision of translation are a "hard no" to implementing this approach currently. This is not only to preserve the scientifically validated and precise information and protocols housed in SAWBO's videos but also because any imprecision and inaccuracies reduce trust in SAWBO's video content and potentially mislead end-recipients in terms of practices. For example, end users might lose beans stored incorrectly in hermetically sealed containers; this both decreases food security and affects future crop planting, family nutrition between planting seasons, and even the availability of additional income if it becomes necessary to sell some beans to meet farm or family costs. Or in the case of the scouting and spraying technique, as another example, to prevent the spread of fall armyworm, incorrect application of the protocol (due to an LLM's slightly mistranslated emphasis in video content) could expose farmers to crop destruction. Worse still, because LLMs assemble their corpus of data by scraping the most available languages of the world, then it will be precisely the most minority, least-spoken languages that will suffer the most in terms of imprecision and inaccuracies.

It is incredibly tempting to hope to leverage automatic LLM translation, because then it is no longer necessary for SAWBO (or anyone) to spend the time, resources, or money to produce a linguistically relevant and comprehensible language variant for any one person that others in the area could use; the SAWBO app would provide it. Better still, the translation generated could then be uploaded to SAWBO's library and become available for all other speakers of that language, even highly minority or endangered languages. But as long as imprecision and ad hoc errors distort the scientific precision and meaning in SAWBO's videos, such an approach is not just infeasible but completely unacceptable. There is also the issue that, ideally, any translation must be voiced over the video's visual content as well; the temptation to automate this adds simply another layer where AI imprecision and inaccuracy unacceptably risk the destruction of the content's integrity, erode faith in the content, and represent a breach of confidence and ethical responsibility between the content provider and any end recipient (Medendorp et al., 2023).

Instead, it may be feasible to accelerate the translation of video content by using AI algorithms and other means to invite qualified language speakers to volunteer translations that they then vet for accuracy and scientific precision. This invitation could streamline the process of widening the net of languages encompassed by the SAWBO library. Another potential advantage of this approach would be that the SAWBO library could begin to accumulate dialect variants. Sometimes "dialect discrimination" can lead to the perception that speakers of certain dialects (whether in major or minority national languages) possess less authority, credibility, or intelligence (Baugh, 1999). How exactly to operationalize or organize such additional variants, and whether they effectively and cost-effectively extend the right of knowledge to still greater numbers of people, would be a useful topic for future research.

Conclusion

This chapter has examined how AI, machine learning, and big data technologies can support the achievement of the UN's Sustainable Development Goals across global, national, and local scales, focusing especially on SAWBO's implementation. Through these tools, SAWBO has shown how to analyze gender dynamics in Bangladesh, reduce video delivery costs, and optimize content distribution via its app—demonstrating the real potential of technology to accelerate access to life-critical knowledge.

When deployed ethically and strategically, these technologies can help address wicked global challenges and establish the groundwork for a universal right to knowledge. SAWBO's use of AI and machine learning has already cut costs dramatically: the app can reach 100,000–200,000 users for the same investment it takes to train a few dozen people in person. This level of scalability marks a major shift in what is possible for ICT-based development interventions.

Yet key challenges remain. Chief among them is the accurate translation of video content into the world's many languages. While large language models offer a tempting shortcut, current imprecisions—especially in low-resource or minority languages—risk distorting scientific content and eroding user trust. Until such tools can ensure precision, SAWBO and similar efforts must rely on more rigorous translation methods, potentially involving qualified local speakers to review and refine AI-assisted output.

Looking ahead, SAWBO could build on its existing app's infrastructure by integrating features like push alerts for new region- or language-specific content and tools to join local communities of practice based on shared training interests. With over 20 WhatsApp groups already operating in Africa and beyond, the foundation exists for a growing ecosystem of collaboration, adaptation, and knowledge sharing. Ultimately, SAWBO's model of cost-effective, scalable knowledge dissemination offers a compelling prototype for how development technologies might move beyond access and into agency—reaching not only more people but doing so in ways that empower them to action. The future lies in refining these tools to meet the needs of diverse linguistic communities, and in designing systems where the right to knowledge is not aspirational, but operational.

References

Alzubi, J., Nayyar, A., & Kumar, A. (2018). Machine learning from theory to algorithms: An overview. *Journal of Physics: Conference Series, 1142*, 012012. https://doi.org/10.1088/1742-6596/1142/1/012012

Anderson, C. (2008). The end of theory: The data deluge makes the scientific method obsolete. *Wired, 16*(7), 16–17.

Bachmann, N., Tripathi, S., Brunner, M., & Jodlbauer, H. (2022). The contribution of data-driven technologies in achieving the sustainable development goals. *Sustainability, 14*(5), 2497. https://doi.org/10.3390/su14052497

Bartley, K. (2024). *Big data statistics: How much data is there in the world?* https://rivery.io/blog/big-data-statistics-how-much-data-is-there-in-the-world/

Baugh, J. (1999). *Out of the mouths of slaves: African American language and educational malpractice*. University of Texas Press. https://books.google.com/books?id=xoZ0POyF2YkC

Bello-Bravo, J., Medendorp, J. W., Lutomia, A. N., Reeves, N. P., Tamò, M., & Pittendrigh, B. R. (2022). Dramatically increased accessibility and decreased cost-per-person impacts are needed for scaling IPM in Africa. *Current Opinion in Insect Science, 54*, 100971. https://doi.org/10.1016/j.cois.2022.100971

Brożek, B., Furman, M., Jakubiec, M., & Kucharzyk, B. (2024). The black box problem revisited. Real and imaginary challenges for automated legal decision making. *Artificial Intelligence and Law, 32*(2), 427–440. https://doi.org/10.1007/s10506-023-09356-9

Cambridge, D., Wenger-Trayner, E., Hammer, P., Reid, P., & Wilson, L. (2024). Theoretical and practical principles for generative AI in communities of practice and social learning. In A. Buch, Y. Lindberg, & T. C. Pargman (Eds.), *Framing futures in postdigital education: Critical concepts for data-driven practices* (pp. 229–239). Springer.

Cukier, K. N., & Mayer-Schoenberg, V. (2013). *The rise of big data: How it's changing the way we think about the world*. https://www.foreignaffairs.com/articles/2013-04-03/rise-big-data

Dyer, R. (1997). *White: Essays on race and culture*. Routledge.

Eghosa, F. (2025). *I tested Grok 3 vs. ChatGPT with these 10 prompts: Here's the winner*. https://techpoint.africa/guide/grok-3-vs-chatgpt/

Fountain, J. E. (2022). The moon, the ghetto and artificial intelligence: Reducing systemic racism in computational algorithms. *Government Information Quarterly, 39*(2), 101645. https://doi.org/10.1016/j.giq.2021.101645

Ghahramani, Z. (2015). Probabilistic machine learning and artificial intelligence. *Nature, 521*(7553), 452–459. https://doi.org/10.1038/nature14541

Gibson, R. B. (2006). Beyond the pillars: Sustainability assessment as a framework for effective integration of social, economic and ecological considerations in significant decision-making. *Journal of Environmental Assessment Policy and Management, 8*(03), 259–280. https://doi.org/10.1142/S1464333206002517

IEAG. (2014). *A world that counts: Mobilising the data revolution for sustainable development*. United Nations.

Jamba, F., & Marambi, S. (2024). Sustainable use of artificial intelligent technology: What is it for the Global South in achieving inclusive development goals? In D. Mhlanga & M. Dzingirai (Eds.), *Fostering long-term sustainable development in Africa: Overcoming poverty, inequality, and unemployment* (pp. 343–374). Palgrave Macmillan.

Leonelli, S. (2021). Data science in times of pan(dem)ic. *Harvard Data Science Review, 3.1*(Winter), 2–30. https://doi.org/10.1162/99608f92.fbb1bdd6

Medendorp, J. W., Reeves, N. P., Sal y Rosas Celi, V. G., Harun-ar-Rashid, M., Krupnik, T. J., Lutomia, A. N., Pittendrigh, B., & Bello-Bravo, J. (2022). Large-scale rollout of information and communication technology-enhanced extension training in Bangladesh demonstrates challenges and opportunities towards inclusive gender participation. *PLoS One, 17*(7), e0270662. https://doi.org/10.1371/journal.pone.0270662

Medendorp, J. W., Bello-Bravo, J., Lutomia, A. N., & Pittendrigh, B. R. (2023). The medium for the missing: Participatory animated media for ethical and effective information dissemination. In J. Bello-Bravo, J. W. Medendorp, A. N. Lutomia, & B. R. Pittendrigh (Eds.), *Gender, digitalization, and resilience in international development: Failing forward* (pp. 80–102). Routledge.

Nazeer, M. Y., & Nadir, M. T. (2024). Data deluge dynamics: Tracing the evolution and ramifications of big data phenomenon. *International Journal of Research and Innovation in Social Science, 8*(5), 2147–2156. https://doi.org/10.47772/IJRISS.2024.805157

Nedungadi, P., Surendran, S., Tang, K.-Y., & Raman, R. (2024). Big data and AI algorithms for sustainable development goals: A topic modeling analysis. *IEEE Access, 23*, 188519–188541. https://doi.org/10.1109/ACCESS.2024.3516500

Ni, J., Shen, K., Chen, Y., Cao, W., & Yang, S. X. (2022). An improved deep network-based scene classification method for self-driving cars. *IEEE Transactions on Instrumentation and Measurement, 71*, 1–14. https://doi.org/10.1109/TIM.2022.3146923

Noble, S. U. (2018). *Algorithms of oppression: How search engines reinforce racism*. NYU Press.

O'Leary, D. E., & Storey, V. C. (2022). Data exhaust. In L. A. Schintler & C. L. McNeely (Eds.), *Encyclopedia of big data* (pp. 283–285). Springer.

Omodei, E., Kim, D., Garcia-Herranz, M., & Sekara, V. (2023). Are machine learning, AI, and big data tools ready to be used for sustainable development? Challenges, and limitations of current approaches. *Frontiers in Big Data, 6*, 1301903. https://doi.org/10.3389/fdata.2023.1301903

Reeves, N. P., Ramadan, A., Sal y Rosas Celi, V. G., Medendorp, J. W., Ar-Rashid, H., Krupnik, T. J., Lutomia, A. N., Bello-Bravo, J. M., & Pittendrigh, B. R. (2023). Machine-supported decision-making to improve agricultural training participation and gender inclusivity. *PLoS One, 18*(5), e0281428. https://doi.org/10.1371/journal.pone.0281428

Ruijer, E. (2021). Designing and implementing data collaboratives: A governance perspective. *Government Information Quarterly, 38*(4), 101612. https://doi.org/10.1016/j.giq.2021.101612

Sahoo, B. A. (2021). The technical buzzword: Artificial intelligence. *Indian Journal of Artificial Intelligence and Neural Networking, 1*(4), 1–3.

Sarker, I. H. (2021). Data science and analytics: An overview from data-driven smart computing, decision-making and applications perspective. *SN Computer Science, 2*(5), 377. https://doi.org/10.1007/s42979-021-00765-8

SAWBO. (2018). *How to identify and scout for Fall Armyworm in English (accent from USA)*. https://www.youtube.com/watch?v=5rxlpXEK5g8&feature=emb_logo&ab_channel=SAWBO%E2%84%A2ScientificAnimationsWithoutBorders

SAWBO. (2025). *Scientific animations without borders video library*. https://sawbo-animations.org/

Stilgoe, J. (2018). Machine learning, social learning and the governance of self-driving cars. *Social Studies of Science, 48*(1), 25–56. https://doi.org/10.1177/0306312717741687

Tang, J., Liu, G., & Pan, Q. (2021). A review on representative swarm intelligence algorithms for solving optimization problems: Applications and trends. *IEEE/CAA Journal of Automatica Sinica, 8*(10), 1627–1643. https://doi.org/10.1109/JAS.2021.1004129

Teh, D., & Rana, T. (2023). The use of internet of things, big data analytics and artificial intelligence for attaining UN'S SDGs. In T. Rana, J. Svanberg, P. Öhman, & A. Lowe (Eds.), *Handbook of big data and analytics in accounting and auditing* (pp. 235–253). Springer.

Tian, J., Xie, H., Hu, S., & Liu, J. (2021). Multidimensional face representation in a deep convolutional neural network reveals the mechanism underlying AI racism. *Frontiers in Computational Neuroscience, 15*, 620281. https://doi.org/10.3389/fncom.2021.620281

van Es, K., & Masson, E. (2018). Big data histories: An introduction. *Tijdschrift voor Mediageschiedenis, 21*(2), 1–6.

Vershynin, R. (2018). *High-dimensional probability: An introduction with applications in data science* (Vol. 47). Cambridge University Press.

Wang, Z., Yan, H., Wei, C., Wang, J., Bo, S., & Xiao, M. (2024, August 9–11). Research on autonomous driving decision-making strategies based deep reinforcement learning. In *Proceedings of the 2024 4th International Conference on Internet of Things and Machine Learning, Nanchang, China*. Association for Computing Machinery.

Xu, H., & Shuttleworth, K. M. J. (2024). Medical artificial intelligence and the black box problem: A view based on the ethical principle of "do no harm". *Intelligent Medicine, 4*(1), 52–57. https://doi.org/10.1016/j.imed.2023.08.001

Zimmerman, A., & Nardi, B. A. (2016). Two approaches to big science: An analysis of LTER and NEON. In J. N. Parker, N. Vermeulen, & B. Penders (Eds.), *Collaboration in the new life sciences* (pp. 65–84). Routledge.

Chapter 6
The Right to Traditional Knowledge and Language

A Framework of Rights

These days, one often hears people loudly proclaiming, "I have the right to such and such," frequently without any legal basis or knowledge of relevant case law. Nonetheless, such assertions reflect a fundamental human insistence on a person's deservingness to participate in public life and to be given due process when that deservingness feels violated. The phrase "I have the right" represents an assertion of personal autonomy and agency, signaling that the individual believes they deserve certain freedoms, protections, or resources integral to their dignity and self-determination. It also serves as a recognition of social norms—whether tacit or codified—that are collectively asserted or accepted broadly by society. By invoking a right, people appeal to these shared values, reinforcing or demanding the norms and expectations that hold society together (c.f., the Diné notion of *k'é*). In this way, the claim of rights—poorly informed or not—is central to negotiation and conflict resolution within the culture, as it reminds all parties of the collective terms of engagement agreed to in principle and provides a basis for dialogue, legal adjudication, or conflict resolution (Bello-Bravo et al., 2022; Sen, 2012).

However, it is critical not to oversimplify the distinction between "formal" and "informal" conflict resolution. While all cultures maintain procedures for settling conflicts—whether they involve harms against others or breaches of agreement—"informal" conflict resolution like mediation and negotiation can be highly intricate and based on precedent like "formal" litigation in court. For instance, restorative justice explicitly offers a formal alternative to punitive litigation in nation states. By focusing on repairing the harm caused by wrongdoing and actively involving the victim, the offender, and the community in a process that promotes healing, accountability, and the restoration of relationships, frameworks of restorative justice draw on, or echo, similar indigenous practices, including Diné *hózh̨óji naat'aah* (Navajo peacemaking: Fletcher, 2014; Lee, 2011), Anishinaabe "circle justice" (Mills et al.,

J. Bello-Bravo et al., *The Right to Knowledge*,
https://doi.org/10.1007/978-3-032-03360-4_6

2017), Māori Marae-based justice (Tauri & Morris, 1997), and Rwandan *gacacas* (Longman, 2009). European traditions antedating the nation-state—including the Þing in early Germanic and Scandinavian societies (Byock, 2001; Imsen, 2013), the Icelandic *Landnámabók* (Pálsson & Edwards, 1972; Stanhope, 2024), Brehon law in pre-Christian Ireland (Kelly, 1988), the *weregild* of Anglo-Saxon and other Germanic tribes (Devi-McGleish & Cox, 2018), and Slavic *veche* (Arakcheev, 2013)—also could more emphasize community consensus and restitution than punishment per se (Zehr, 1990).

This distinction between nation states and other forms of conflict resolution is significant because, while there may be as many as 5000 sovereign First Nations globally—including 1674 formally recognized indigenous peoples in Canada, Australia, and the United States—the United Nations officially recognizes only 193 countries (and two observer states, the Vatican and Palestine). As a result, frameworks around "rights" are predominantly defined and established by nation states, and it is those frameworks that are usually, if not exclusively, recognized between governments. Solely for this reason, we discuss the right to know through a nation-state lens in this chapter, but it remains important to acknowledge their overall differences with respect to rights in other traditions.

In nation states, a "right" is usually understood as a legal entitlement granted and protected by a government's laws and constitution (Malesevic, 2013; Sanya, 2017). These rights are codified, formalized, and often subject to interpretation and enforcement by institutions, such as courts. Nation states emphasize individual rights, which are framed within a legalistic structure that allows for adjudication, litigation, and state enforcement. Rights in this context are universal within the government's jurisdiction, applying equally in principle (and less often in practice) to all citizens or subjects. Despite backing by the government, they are often construed as inherent or inalienable, particularly where civil liberties and human rights are given a premium.

Paradoxically, national and local governments sometimes pose a roadblock to enabling low-literate learners to access knowledge. Such roadblocks may range from the lack of resources to enable the sharing of this knowledge ("passive roadblocks") to actively attempting to block people's access to knowledge ("active roadblocks") (Payumo et al., 2021). Several of this book's co-authors have sat in the office of an official who demanded "resources" in exchange for "allowing" farmers to obtain access to knowledge—spending over one year blocking farmers' access to this content—setting an active and intentional obstacle. SAWBO's app described in Chap. 5 allowed local entities to access and use materials as they saw fit, even while the official maintained for more than a year that farmers should not access this content until they (the official) received "resources" and gave the content official sanction (despite not being a content-area expert on the videos' topics). Notably, content experts within the same organization had given their prior approval of the content before the official became involved. Also, no "resources" were ever provided by SAWBO, and after one year, the official finally relented and decided to "review" the videos to give approval. That review of the several 3- to 6-min-long animations took several months, at the end of which, the videos were deemed "fit as is" for farmers to use.

The authors would like to clarify that this incident was a onetime experience. As a general rule, government agencies have actively worked with SAWBO or engaged in very positive collaborations (Lutomia et al., 2020). Exceptions to a rule can provide important learning experiences. In other research (Bello-Bravo et al., 2021), we notice how the development of globally good e-extension materials—placed into a format that is both visually and auditorily understandable by local populations, along with platforms to facilitate easy access and use—holds the potential to circumvent both passive and active roadblocks. Most importantly, for governments, government officials, NGOs, and all other entities and change agents wishing to share such educational content, by providing it freely for use, SAWBO bypasses the "passive" roadblocks of limited resources and sociopolitical barriers.

While cases of "active roadblocks" are murkier, by nonetheless making content freely available for anyone to use as they see fit, it becomes more challenging for those elites who are intent on blocking people's right to knowledge to succeed. This obstruction includes the example (described in Chap. 3) of the minority language activist whose national government refused to provide COVID-19 prevention information in their community's mother-tongue. But one may also reasonably infer that in places where women's access to knowledge may be socially (not just economically) constrained, knowledge access via the app or SAWBO's downloads can again avert that constraint (Medendorp et al., 2022). Also, digital portals that can be accessed anonymously, like WhatsApp, may also circumvent denials of women's access to knowledge (Lutomia et al., 2024) and their practice of its right. In this way, we consider SAWBO's lack of control over who downloads and uses its content as an asset not a liability, besides that we also feel no desire to control who uses our content for educational purposes.

If one turns from government contexts to indigenous/traditional contexts, a "right" is often more closely tied to communal responsibilities and relationships rather than individual entitlements (Gibson, 2008; McDonald, 1998). Rights in these societies remain embedded within a broader cultural, spiritual, and social framework, where the focus is on maintaining social harmony, balance, and the well-being of the community as a whole. These rights are not necessarily codified in written laws but are passed down through oral traditions, customs, and practices. They become deeply connected to the land, identity, and the collective life of a people. While sometimes relying on formal legal systems for enforcement, rights are often upheld through communal consensus, conventional leadership, and restorative practices that aim to restore relationships and repair harm more than punish wrongdoing. Such rights often intertwine with duties to the community, the environment, and future generations, reflecting a more sustainable and integrated approach to justice and governance. How different this sense of rights and conflict resolution is from nation-state approaches is evident in traditional Diné [Navajo] Peacemaking. Drawing on the traditional Journey Narrative, by which the Holy People traveled through the four worlds:

> [Peacemaking] begins in a place of chaos, *hóóchx̨o’/anáhóót’i’*, whether within an individual or between human beings. Perhaps due to historical trauma, Navajos shy away from face-to-face confrontations. However, such confrontations are vital in order to dispel [chaos]. The Peacemaker has the courage and skills to provide the groundwork for the person or group to confront *hóóchx̨o’/anáhóót’i’* and move toward mastering harmonious existence. Life-value engagement with the Peacemaker provides the sense of identity and pride from our cultural foundations. *Hóóchx̨o’/anáhóót’i’* can block and overwhelm clanship, *k’é*, which is normally what binds human beings together in mutual respect. Through engagement, the Peacemaker educates, scolds, persuades, pleads, and cajoles the individual or group toward a readiness to open up, listen, share, and make decisions as a single unit using *k’é*. When *hóóchx̨o’/anáhóót’i’* is confronted, people may learn there is a choice to leave it. When harmony, *hózh̨ó*, is self-realized, sustaining it will have clarity and permanent *hózh̨ó* will be self-attainable, *hózh̨ǫ́ójí k’ehgo nįná’íldee’ iłháháhodidzaa ná’oodzíí’*. Through stories and teachings, the Peacemaker dispenses knowledge, *naat’áánii,* in order to guide the whole toward a cathartic understanding of *hózh̨ó* that opens the door to transformative healing. The flow of *hózh̨ó* is a movement inward toward the core issue or underlying truth. Recognition of this truth and the ending of denial provide the opportunity for healing or mutual mending. Realization of the truth occurs when individual feelings are fundamentally satisfied. The resolution of damaged feelings is the core material of peacemaking sessions, *hózh̨óji naat’aah*. Depending on the skill of the Peacemaker, *hózh̨ó* may be short or may take several peacemaking sessions (NSN, 2024).

The Right to Knowledge

We must acknowledge that, with respect to any *right to knowledge* per se, no universally recognized legal framework as yet exists that formally establishes or protects such a right. Nonetheless, various proxies can be cited that suggest or approximate it. For example, a myriad “right to know” legislations and rules exist globally that mandate safety notifications about potentially hazardous conditions, e.g., in the workplace, about public groundwater contamination, or as labeling for hazardous materials and GMOs. Broadly, UNESCO frames the right to know as a human right to “freedom of information” such that people can “participate in an informed way in decisions that affect them, while also holding governments and others accountable” (UNESCO, 2010, p. 16); this definition includes resisting limitations on journalistic freedom, overcoming technological problems of access, and increasing stakeholders’ participation in all sectors, including international development (Cullen & Coryn, 2011).

Clearly, such “freedom of information” can be a “right to knowledge” *if* that information does in fact empower people to participate in an informed way, not only with respect to decisions that affect them, but toward solving problems they perceive in their world, community, or family. This empowerment recalls Chap. 1’s framing of knowledge as the *application of* or *enabling of* a course of action (Nonaka & Takeuchi, 1997). Through the distinction between the *reach* of an information campaign and recipients’ *engagement* with it (Johnson et al., 2018), one might suggest that “information” is transmitted but “knowledge” is (ideally) received. Once that new capacity for action is actually put to use, then we can further understand the manifest changes in the world around the person as evidence of *empowerment*, "an expansion in people's ability to make strategic life choices, in a context where this ability was previously denied to them" (Kabeer, 1999, p. 437).

For this reason, we underscore throughout this book people's *exercise* of their right to knowledge as empowering. Without being able to put knowledge to use, it is merely information. This distinction may seem an overly fine hair to split, but for international development work and humanitarianism overall, there is an ever-present danger of believing that merely "getting the message out" (maximizing reach) suffices (Bello-Bravo, 2020). Partly this hazard is a logistical constraint, given that maximally up-scaling reach is often the only funded budget aspect of a project (Reeves et al., 2020); partly the problem is methodological, given that measuring the actual engagement (and so also the empowered use) of such maximally up-scaled information dissemination is difficult (and thus rarely funded as part of the project) (Bello-Bravo et al., 2013). But the problem is also partly one of attitude (Daré et al., 2014). Certainly, to go to the trouble of disseminating helpful and life-sustaining information to people, only for them to not receive it or take it up can be dispiriting and frustrating. But the analysis of why these shortfalls and failures occur too often focuses on characteristics of the recipient rather than the adequacy of the message or media (Bello-Bravo et al., 2017, 2023a).

Plainly enough, it is contradictory to fault end users for failing to take up a solution transmitted to them in a language they cannot comprehend, through a channel they cannot access, or at times and places where it is prohibitively inconvenient or socially forbidden to do so. We might instead remember our own "failures" to take up similarly "irrelevant" or "inconvenient" solutions. In that way, it is not a mystery that others would reject a solution that is really not a solution at all and even less so any "evidence" that they are not interested in improving their situation or for some reason are irrationally refusing help.

As unpleasant as it can be to discover that what we offer as help is not up to the task, we must still guard against the temptation "to attribute primary responsibility [for that failure] to disagreeable or damaged 'others' whose bad acts are thought to be the products of their flawed characters and ill-advised choices" (Haney & Zimbardo, 2009, p. 807), whether or not we acknowledge people's situational constraints impacting their actions (Gawronski, 2004). Overcoming the logistical and methodological challenges of delivering conveniently accessible and practically feasible solutions that end-recipients actually embrace may be difficult; changing our attitude toward who we help and how we help them is not.

In practical applications, it is incumbent on solution providers to intentionally dialogue with end-recipients before determining what solutions might be appropriate—and better still, to urge recipients to take the lead in determining what is needed. As the then-Minister of Education from South Sudan said at the 2015 Higher Education Conference in Dakar, Senegal, "First they told us that we were needy. Then they told us what we needed" (as heard by one of our co-authors in attendance at the conference). This comment was uttered tongue-in-cheek, of course, but the gist of the remark was unmistakable: *how can we, as indigenous peoples, trust the solutions offered us when we have no say in defining the process?*

SAWBO remains cognizant of these delicate balances between external solutions and local needs (Bello-Bravo et al., 2020b). For that reason, content generation is always a collaborative process. Recipient groups are encouraged to participate in the definition of the problem as well as the solution. This point is not to imply that the scientific value of a solution is somehow mitigated—only that engagement with local users helps shape the issues for which solutions might be appropriate (Bello-Bravo & Amoa-Mensa, 2019); from there, one can then frame solutions such that they address a genuinely felt need in the community.

In this way, the right to knowledge is not some abstract construct that simply dumps external solutions on local users. It is a right to knowledge that is relevant, contextualized, locally defined, and applicable; this scope makes it an ethical responsibility for solution developers to provide (Medendorp et al., 2023). This obligation requires a larger enterprise than simply aggregating the world's knowledge and putting it at people's doorsteps in the hope that they take it up. Rather, it involves an intentional and designed effort to deliver the right content to the right people, at the right times and places, and in the right way so that it is useful to local populations. Supporting the right to knowledge means ensuring that knowledge addresses real felt needs in ways that are immediately applicable to local contexts.

Antecedents to a Right to Knowledge

While "freedom of information" is the indispensable prerequisite to putting that information to use as exercise of the right to knowledge, concepts around the right itself draw on other frameworks of rights to support accessing information. Above all, Article 26 of the *Universal Declaration of Human Rights* and Article 13 of the *International Covenant on Economic, Social and Cultural Rights* emphasize a right to formal education. To the extent that education provides access to information and hence the potential for knowledge's empowerment, a right to formal education may seem to automatically entail a right to knowledge and thus settle the matter. However, a considerable literature critiques the operations of formal educational systems. As Reynolds (2021) stresses, ideology and its accompanying social ignorance can be "understood not simply as the absence of education, but as *a problematic form of consciousness produced and maintained through educational institutions*" (p. vi, emphasis added); again, these entities are primarily formal educational institutions. Similarly, Heydorn (2024) observes the form(s) of education that became integral to the European Enlightenment exhibit "a double necessity: to equip humankind for the revolution of the productive forces and to prevent the revolution of one's consciousness" (p. 10); thus, education simultaneously produces and prevents certain forms of knowledge. In this way, education's "progressive liberation of humanity to itself … now becomes the most developed instrument to close off the knowledge of one's own possibility to humanity" (p. 10).

We note these critiques do not yet propose a moratorium on formal education (Illich, 1971) but require mindfulness of the consequences that result from formal educational systems (Alexander, 2012; Brewster, 2020; Freire, 1970; Roediger, 2019; Skiba et al., 2014). Since the goal of this book encompasses describing how to best secure the exercise of the right to knowledge overall, this objective demands an availability, usability, and convenience of information systems' access for all people (youth and adults alike), whether through formal education venues or informal and non-formal learning (Souto-Otero, 2021; Tudor, 2013), whether in short-term and lifelong learning contexts (Avoseh, 2010; Lutomia & Bello-Bravo, 2024).

As such, proposals to open up existing formal educational systems to those people and groups who have been, and are still being, denied access to them are a step forward only if the curricular information provided by them is convenient and applicable for the realization of their human possibility (Bello-Bravo & Lutomia, 2024). This provision becomes mandatory in adult education (Knowles, 1990), where adult learners typically seek further learning with specific goals in mind, e.g., skill acquisition, curiosity about a topic, personal growth (Anyanwu & Pappoe, 1981); failing to address that practical interest often results in attrition and non-participation (Mezirow, 1991).

To the extent that a right to adequately available, usable, and convenient education also supports a right to knowledge, other frameworks can further add to the picture. Article 19 of the *Universal Declaration of Human Rights* also affirms a right to freedom of expression—especially as a right to dissidence—which positions such expressions as critical social information that people hold a right to be aware of. Similarly, legislation like the U.S. *Freedom of Information Act* outlines guidelines for accessing otherwise non-disclosed information, while intellectual property laws—to the extent that they establish procedures for what and how one may use information—are part of that discussion as well. Truth and reconciliation commissions represent another critical embodiment of the right to knowledge (Llewellyn & Howse, 1999). So too the many rights, limitations, and laws concerning privacy and confidentiality, including the right to be forgotten in an age of mass data breaches and emergent facial recognition technologies, predictive policing, unmonitored decision-making by expert systems, and Big Data aggregation generally (Fabbrini & Celeste, 2020; Jefferson, 2020; Memarian & Doleck, 2023; Williamson & Prybutok, 2024).

As predominantly *de jure* articulations around data, information, and knowledge, these additional frameworks delimit and define more than expand exercising the right to knowledge. In contrast, "open-access" movements have offered pushback and critiques of specific legislation (and *de jure* approaches overall), especially where denials of access to data or information are perceived either as not convincingly enough framed in the public's genuine interest or too much in the interests of private, entrenched parties (Peters et al., 2007). Public libraries also have long been sites where staunch advocacy for broad (usually unlimited) access to information and the protection of individual library patrons' privacy are core principles (Dresang, 2006; Popowich, 2020). Also, Columbia University's celebrated 1954 bicentennial radio lecture series "Man's Right to Knowledge and Its Free Use Thereof" eventually reached a global-level of influence and "helped to give to millions of individuals a deeper understanding of their rights to knowledge" (Peabody Awards, 1954).

While these de jure frameworks and de facto movements suggest a basis for a right to knowledge, a brief excursion through the concept of a right itself is in order. What is usually most missing from those loud, out-of-the-blue assertions of *I have the right to* such and such is any awareness that invocations of a *right* come indissolubly with the responsibilities associated with that right—otherwise, it's not a right; it's merely license. It was in principle to put a check on such self-assertive license by monarchs and aristocrats—even if it cleared the way for the same by an emergent bourgeoisie—that Enlightenment thinkers, principally Hobbes, Locke, and Rousseau, were motivated to reformulate the social contract (Rawls, 1971; Williams, 1980). They contended—often on moral grounds pulled quite out of the thin air—for a recognition of certain *necessary* rights and responsibilities inhering between individual members of a State and the State itself. While the details of this social contract varied, subsequent historical developments gradually inscribed all manner of such rights and responsibilities into legal codes, lease agreements, EULAs, student handbooks, terms of service, patient consent forms, codes of conduct, privacy policies, and contracts of all sorts. In all of these cases, a remedy for breach is also stipulated, whether or not a party can pursue it or not.

Presently, this legalistic framework of rights is so deeply entrenched in the mechanisms of its habitual practice that one forgets its originally *moral* basis of argument—as a brash and desperate special pleading ostensibly propped up by Reason, that the State and other powerful institutions *should* respect the autonomy, property, and being of its individual members. While it was the responsibility of individual members of a State to "consent" in various ways to the State's protection and its processes to ensure the continuance of collective society, the State in return held the responsibility not to dispossess, cripple, or otherwise render individual members' lives impossible—especially not to seize the fruits of emergent bourgeois enterprise, property, or land.

In this sense, the historical circumstance of slavery at the time the U.S. *Declaration of Independence* throws a sharp light on what *inalienable* rights claimed by the British colonies to life, liberty, and property (the pursuit of happiness) looked like—rights, they insisted, that were being flouted by the Crown. The strength (not to say the irony) of this moral insistence is that it makes *recognition* of one's autonomy, property, and being non-contingent on any whim or rationale of a State; it also means such rights cannot be rescinded by fiat or legislation.

That *inalienable* rights might still be ignored or violated could still occur—as indeed the Colonies were specifically complaining of the Crown—but any such violations also could, and indeed *should*, be viewed as flagrantly irrational (and immoral) by any rational person. The weakness or risk of this argument is that it depends on the vulnerability of an entrenched interest to feel *stung* by such an accusation; a sufficiently shameless or insulated tyrant will remain unmoved by such a charge, at least publicly. Thus, whether George III of England felt morally stung by the Colonies' aspersions toward him or not, he responded with troops sent to make a strong show of putting up with no such guff and putting down an insurrection, albeit unsuccessfully.

This vulnerability of Power (and the vanity of people invested with it) to feeling stung, to being liable to *moral* accusations is a curious and important chink in the often seemingly impenetrable armor of Power, especially in a postmodern situation of learned political helplessness (Courington, 2005). A recent example of this vulnerability involves the move (whether genuine or merely for show) toward "green" and corporate responsibility (De Vries et al., 2015; Pjevovic, 2017; Watson, 2017). This vulnerability of an otherwise generally indifferent and unassailable organization to the threat of negative public *perception* discloses an essential arrow in the quiver of moral, practical, and reasonable arguments for a right to knowledge—again invoked almost entirely out of thin air—especially for those people who have most-been disenfranchised, marginalized, erased, and ignored historically.

Nevertheless, being vulnerable to such arrows requires a fundamental agreement on the validity of the rights in question. Otherwise, accusations and protests bear no power. Here then, we notice again the power of knowledge, both as a weapon and as a tool. On the one hand, the knowledge that is shared can reinforce power structures, communicating to each member of society what their "role" should be. On the other hand, when knowledge seeps out and undermines the official narrative, it empowers those people who receive it to question the status quo and its assigned roles.

For this reason, it is essential to use tools that carry a power to circumvent the official channels of communication and to speak directly to end users. Although some countries have invested great sums to control the flow of information, ICT-based communications offer a way of leaking out and finding those folks in need. Although nothing in SAWBO's inventory is revolutionary in the sense that it directly disrupts power structures, providing knowledge empowers people not only intellectually but also economically, with a resultant incremental control over one's own destiny. In that sense, SAWBO's goals coincide with the ethos of the SDGs: to empower the people of the world with the freedom to choose their own destiny by providing them with the basics of life (Kabeer, 1999; Kjosavik, 2021).

The Right to Traditional Knowledge

Logistically, to exercise a right to knowledge depends on access to widely available, wholly usable, and readily convenient information. Thus, the technical aspects addressed here describe the several what, where, and when questions around information accessibility; the social questions of who, how, and why remain challenges.

While we have already referred in Chap. 1 to (digital) communities of practice accessing knowledge (at all scales from individual change agents to global scale international partnerships), more needs to be said about *which* knowledge, beyond simply emphasizing gold-standard, life-sustaining knowledge for security around food, water, health, shelter, and gender as critical components of the SDGs. Knowledge is not only (or ever) an immaterial abstraction; rather, it appears in material ways, as documented media and living memory in people. If we then imagine something like a global public digital library or an encyclopedia containing all life-sustaining knowledge online, then we can ask not only what media are available to be checked out in that library or read in the encyclopedia but also what items are missing.

Anthropologist Paul Sillitoe (2002) reported two decades ago of regularly encountering claims within international development spaces about the presumed illegitimacy or anthropological delusion of the very *concept* (never mind the actuality) of indigenous knowledge. To whatever extent things have since changed (for the better or the worse), we must not lose sight of how *casually* some people could voice outright denials of the life-ways and experiences of peoples still subjected to systematic social erasure (Robertson, 2015). Without intending any other historical analogies except to underscore this casual tone, one can also note that anti-woman backlash retains a still-current and broad license globally (Dragiewicz, 2013; Van Wormer, 2008; Yeung et al., 2014). We must consider this situation the first major barrier to an exercise of the right to indigenous and women's knowledge: a prevailing denial of their legitimacy.

Incommensurable differences hold between the knowledge and practices of indigenous people and non-indigenous women across time, space, and historical setting. We do not aim to glibly or unselfconsciously analogize these differences, though one can trace the efforts of others who did—particularly an Enlightenment patriarchy that fixed women and the indigenous spatially and conceptually in the home and on the reservation (Ellis, 1989; Jackson, 2024; Morgensen, 2011). Historian Gerda Lerner (1986) characterized patriarchy as a varying but universal inheritance of human culture (indigenous and non-indigenous alike), and Riane Eisler (1994) also contrasts women's horizontal *caring* economy with men's vertical *dominator* paradigm. The example of early feminist women in the United States learning from indigenous (Haudenosaunee) women's more equitable social status is sustaining, ironic, and instructive (Wagner, 2011).

Let us reinforce the point here that some forms of humanity's oldest knowledge can still be discerned around the hearth, the cook pot, the loom, in indigenous women's processing of shea butter, modernized production of textiles, and above all childcare. Even in the reimagining of the bourgeois home that accompanied the spread of the Industrial Revolution(s), women's role as a domestic angel in the home was so exalted that it became an argument to confine them there and deny them access to knowledge (Ellis, 1989).

Of course, it is not that men offer no traditional or updated-traditional knowledge (inside or outside of indigenous cultures, in the past or the present); only that when "knowledge" is invoked in a broad way, one can notice an emphasis on activities outside of the home; a tendency that makes less visible the knowledge and practices relevant to what scholars identify in the unpaid labor of social reproduction by women (Bakker, 2007). To the extent that the biological reproduction of the species is a necessary core of social reproduction overall, the thousands of indigenous and women's traditional and contemporary knowledge(s) around what is referred to broadly as midwifery (Bello-Bravo, 2023) embody an equally core part of life-sustaining knowledge itself (Bello-Bravo, 2020). Other parts of this economy—in the original Greek sense of "household management"—like cooking, sewing, laundering, gardening, and childrearing in all of its facets are, again, literally life-sustaining.

Language represents a specifically indigenous characteristic of knowledge. That Humphries et al. (2013) must advocate for "the existence of a state constitutional legal right to language" (p. 872) offers evidence that the right either does not formally exist or only inadequately at so present. While U.S. federal courts have extensively ruled in favor of "a constitutionally protected right to express oneself or receive communications in a language other than English" (Piatt, 1986, p. 885), the situation that Moria Paz (2013) highlights in an international context around the failed promise of language rights applies overall, despite being currently three years into the International Decade of Indigenous Languages (UNESCO, 2022).

Paz (2013) documents the disparity between international commitments "to protect language rights absolutely" and judicial rulings that have not only "favored linguistic assimilation" but also treat "minority language not as a valuable cultural asset worthy of perpetual legal protection, but as a temporary obstacle that individuals must overcome in order to participate in society" (p. 157). Despite rhetoric and de jure legislation to respect a right to language, Paz (2013) indicates how that right operates more to the benefit of a majority-language hegemony or does not suffice in its remedies of de facto wrongs; this point echoes U.S. federal rulings, where "understood" communications (in one's own language) constitute both a humane and fair way of proceeding without being necessarily constitutionally guaranteed, and yet still serve as a person's legal *acknowledgment* of the information, e.g., the presentation of *Miranda* rights in a language that the arrestee understands (Piatt, 1986).

As with knowledge, it is the practice or use of (indigenous) language that presents the crux of the issue in these cases. Bill Piatt (1986) sympathetically surveys a range of contradictory, sometimes humanly decent, other times grossly prejudicial, Court rulings about a non-English-speaker's right to use, speak, and be addressed in a language other than the hegemonic majority's (English). He poignantly highlights a Tenth Circuit decision that acknowledged "the damage suffered by children whose language rights are not respected" (Piatt, 1986, p. 888): hence, "Teaching the [non-English]-speaking child exclusively in English communicates a powerful message to the child that he or she is a second-class citizen" (p. 888, fn 21), that "their language and culture [are] totally rejected" (p. 888).

If the increasingly acknowledged sense that "language is culture" (and "culture is language") (Gao, 2006; Ortiz et al., 2020) can seem an overstatement of the differentiable and homologous but complex interrelatedness of language and culture (Baghzou, 2021; Brock-Utne, 2005), this relationship seems still vividly drawn in the idea that language is "culture expressing itself in sound" (Ovando, 1990, p. 341). Further, because exact distinctions between cultures are difficult to draw—both by the people who live that difference and those anthropologists who study it—language often provides one boundary where those differences between cultures might be drawn.

For indigenous people throughout the majority of human history, their relationship to their language(s) has played a vital role. But, after being forcibly separated from their collective lands and ways of life by dispossession, genocide, and assimilation, many indigenous communities find that language is the only remaining connection to their original culture. There is a pressing need then to preserve, if not restore, the ~40% of languages faced with extinction (most of them indigenous, many without formal print systems). This task is not only to avoid repeating the actionable harm of communicating to their speakers that they are second-class human beings whose culture and language are totally rejected; diversity of language itself is beneficial, if not necessary (Church-Morel & Bartel-Radic, 2016; Evans & Levinson, 2009).

Linguistic diversity benefits from analogies to biology and genetics, where diversity is understood to enhance shock-resilience, adaptive capacity, and the formation of complex, stabilizing relationships within ecosystems. These same dynamics apply to human cultures: Africa's unparalleled diversity of language and tradition—still the richest in the world—offered the fertile ground from which humanity first emerged. But the value lies not just in variety for its own sake. Cultural diversity sustains ethical frameworks that are both adaptive and expansive, including a "live and let live" ethos extended to all forms of life (Baynes-Rock, 2012; Gathogo, 2008; Mangena, 2013) and relational values such as *Bulala* and *Ubuntu*, the latter affirming *I am because others are* (Dolamo, 2013; Lutomia et al., 2018).

Since languages are cultures expressing their perspectives of the world in sound, the edge formed by a multiplicity of those perspectives coming together offers a better promise of emergent insights into the planet's mass-problems than single-voiced and single-languaged echo chambers. Just as monoculture agriculture can increase short-term economic efficiency at the expense of long-term environmental health and sustainability (Corntassel, 2008), the spreading monoculture of English globally may also increase short-term economic efficiency at the expense of long-term impoverishment of fertile ground for creative problem-solving and diminished capacities for social responsiveness, resilience, and stability of the (political) surroundings during crises. Practically and ethically, the dual imperative of preserving indigenous languages and supporting linguistic diversity largely has been a fundamental driving force in SAWBO's strategy to deliver content in as many languages as possible.

Exactly for this reason, (digital) communities of practice—tied to grassroots movements, supported by the government, or leveraged by individual change agents—remain central to this book's approach to learning-systems for achieving the SDGs. Communities of practice can generate, vet, and share insights about a phenomenon. Recognizing a community's cultural wealth of knowledge (Yosso, 2005) represents a critical step in this direction, but communities of practice take it further, by deliberately supporting experimentation and exploration of local experiences about a given phenomenon in order to best understand them. These groups may be not only deft practitioners of indigenous and local knowledge but *theorists* of it as well. This capacity speaks directly to gold-standards of knowledge for building partnerships (SDG17) with local practitioners, which learning-systems can then readily access and move that (local) "basic research" onto the table of solutions on offer.

Digital Communities of Practice

We maintain a critical distinction here. On the one hand, the use of front-end, ad hoc gatherings of local practitioners to explore possible solutions to already-recognized problems (e.g., in focus groups or similar) is assuredly an improvement over providers coming in with a prefabricated single solution and fingers crossed that anyone will want it. Nonetheless, interfacing with a community of practice's experimental engagement with local knowledge offers critical differences. By definition, in a focus group, the multiple overlapping perspectives of practitioners generate an edge and creative brainstorming of insights that would otherwise not have arisen from any one participant alone; however, this creativity is also (very likely) the first time the members of that group have interacted. In contrast, a community of practice forms a history of such interactions, where it is not simply the overlapping perspectives of its members generating insights but temporal edge of its past and present historical domains of practice. Such communities of practices can compare (or may have already compared) existing local/indigenous techniques and any external solutions proffered.

It is crucial to emphasize that communities of practice do not require or depend on an elite group of "experts"; rather, they are grounded by their members in a deep familiarity and local/global engagement with the phenomenon in question—a characteristic often associated with indigenous knowledge and ecological practices (Waller & Reo, 2018). In these ways, a community of practice already holds a capacity to speak authoritatively and convincingly on that phenomenon. In addition to producing credible, accurate information, communities of practice also can serve as catalysts for lowering the transaction costs not only around solution adoption but also solution identification and development (Bello-Bravo et al., 2023c). Indeed, they can *bidirectionally* "translate" necessary behavior changes in culturally legible and feasible ways (Bello-Bravo et al., 2023b).

This importance cannot be overstated, though it is often mis-emphasized. A solution that is technically excellent yet culturally unworkable is either not truly excellent or requires a drastic transformation of the culture for adoption. The latter introduces serious complications, particularly in contexts where Western knowledge systems marginalize or discredit non-Western ways of knowing (Bello-Bravo, 2020; Gandugade et al., 2017; Lutomia, 2019; Nemutandani et al., 2016). This premise strikes the crux of the matter: the competencies of local communities of practice may not align with what Western knowledge systems define as competence, even as those competencies prove effective. The mis-emphasis lies in assuming that such workable knowledge will only be recognized as valid when it is translated into terms familiar to Western science. Here again, the critical role of local communities of practice is their ability to "translate" between Western and non-Western knowledge practices. Equally important is the role of learning systems—especially those that focus on genuine partnerships for accomplishing the SDGs (SDG 17)—which can more readily adopt and adapt insights from communities of practice in the research, development, and deployment of solutions. This bidirectional emphasis on learning by both parties—adapting and adopting knowledge from both sides—is key.

Digital communities of practice, such as those groups empowered by tools like SAWBO's app and WhatsApp SAWBO (and other) groups, represent an ideal model for supporting the right to knowledge. For example, the SAWBO app allows users to access content and move it seamlessly into WhatsApp digital communities of practice, such that that that digital community can explore in what ways the content is useful (or not) in their physical community in a time and place that works for them (and in relation to their local knowledge and practices). These communities not only can download, implement, and benefit from the global encyclopedia of gold-standard knowledge, but their empirical use of this knowledge in local contexts also bears the potential to directly contribute verified discoveries back into the global knowledge base for others to use. A case of this capability was seen during a 2-year follow-up to assess farmers' knowledge retention (93%) and solution adoption (89%) of the postharvest prevention jerrycan technique (Bello-Bravo et al., 2020a). The study found that one in five farmers had not followed at least one of the eight steps of the technique's protocol, while one in three had added an element—specifically, the local tradition of adding chili as a safeguard against insect herbivory. As none of these innovations resulted in failure of the storage protocol (Bello-Bravo et al., 2020a), this example suggests the possibility of locally hypothesis testing whether these modifications—which make the protocol more attractive, trustworthy, and convenient to adopt—could be incorporated into the protocol overall. In a time when international development is in free-fall, the ability of local digital communities of practice to produce knowledge and connect with global-level actors is very much the kind of pivot needed to continue the pursuit of the SDGs.

This digital community of practice approach also leverages the already-detailed ability of SAWBO's app and WhatsApp SAWBO group users to access local modifications to content, particularly linguistic content, thus fostering a more democratized and impactful form of "citizen science." Unlike conventional citizen science, which often involves knowledge producers simply "farming out" data collection tasks, this model supports active experimentation, implementation, and the investigation of gold-standard knowledge in specific locations. It taps into, and empowers, local knowledge experts to share their expertise not only "horizontally" with nearby communities but also "vertically" with global networks that they might otherwise not reach. It transforms receivers of potential information to potential producers of information, from students to teachers—something that achieves its most meaningful consequences when the learning-system of the teacher is willing to become a student itself.

Conclusion

In this chapter, we reflect on the foundations of the right to knowledge in basic human rights and indicate its importance for the achievement of the SDGs. Part of that right is the right to knowledge in one's place, language, and culture. This ideal might be a heavy lift for the international development community, but if we are to accomplish the goal of well-being and ease for everyone through development, then we must consider what that means for the approximately 7000 languages spoken on Earth. Language is the lynchpin to knowledge and knowledge is the lynchpin to freedom.

Knowledge and language remain so closely intertwined that communicating in a local language is an act of respect, a sincere regard for persons and cultures. If we are to achieve the right to knowledge, it is intricately intertwined with the need to honor the languages of the world and, far from obliterating them, to laud and preserve them. If we are to know the "truth," it will require considering a diversity of perspectives, and those perspectives include those people who have been excluded from the "progress" of modern society. ICT-based knowledge bears the potential to both obliterate and to validate. It makes it possible for dominating languages and cultures to overwhelm minority communities, but it also holds the power to preserve those ethnicities that might otherwise disappear. From our perspective, enabling all peoples to contribute to the search for "truth" through knowledge is the only way that we can expect to honor our commitment to improving the human condition as articulated in the SDGs. The challenge and opportunity is that we need to develop pragmatic and actionable ways to achieve this objective.

References

Alexander, M. (2012). *The new Jim Crow: Mass incarceration in the age of colorblindness*. The New Press.

Anyanwu, C. N., & Pappoe, K. A. (1981). *Principles and practice of adult education and community development*. Abiprint Publishing Company.

Arakcheev, V. (2013). Political organization in the Pskov Republic: The prince, the veche and sovereignty. *Quaestio Rossica, 1*(1), 101–116.

Avoseh, M. B. (2010). Informal community learning in traditional Africa. In S. Jackson (Ed.), *Innovations in lifelong learning: Critical perspectives on diversity, participation and vocational learning* (pp. 34–48). Routledge.

Baghzou, S. (2021, March 30–31). Language, culture and related disciplines. In *IALC21: The first international applied linguistics conference: Improving reality through the lens of linguistic theory*, Ouargla, Algeria.

Bakker, I. (2007). Social reproduction and the constitution of a gendered political economy. *New Political Economy, 12*(4), 541–556. https://doi.org/10.1080/13563460701661561

Baynes-Rock, M. (2012). *Hyenas like us: Social relations with an urban carnivore in Harar, Ethiopia* [Dissertation, Macquarie University]. Sydney.

Bello-Bravo, J. (2020). Getting the message across: Characterizing a need to bridge public health messaging for tuberculosis across a rural/urban and CHW/traditional healer divide in Madagascar (a review). *Scientific Africa, 8*(July 2020), e00321. https://doi.org/10.1016/j.sciaf.2020.e00321

Bello-Bravo, J. (2023). Preserving tradition in the here and now: Barriers to the preservation and continuity of traditional healing knowledge and practices in Madagascar. *Social Sciences & Humanities Open, 8*(1), 100696. https://doi.org/10.1016/j.ssaho.2023.100696

Bello-Bravo, J., & Amoa-Mensa, S. (2019). Scaffolding entrepreneurship: A local SME-NGO partnership to enable cocoa production in Ghana. *Journal of Small Business & Entrepreneurship, 31*(4), 297–321. https://doi.org/10.1080/08276331.2018.1528138

Bello-Bravo, J., & Lutomia, A. N. (2024). Knowledge-building, communities of practice, and individuation: Heinz Joachim Heydorn's philosophy of education in relation to an adult learning scientific program in Kenya. In H. Kminek (Ed.), *Survival through Bildung: On the topicality of Heinz-Joachim Heydorn's philosophy of education* (pp. 115–150). Verlag Barbara Budrich.

Bello-Bravo, J., Olana, G. W., Enyadne, L. G., & Pittendrigh, B. R. (2013). Scientific Animations Without Borders and communities of practice: Promotion and pilot deployment of educational materials for low-literate learners around Adama (Ethiopia) by Adama Science and Technology University. *The Electronic Journal of Information Systems in Developing Countries, 56*(1), 1–11. https://doi.org/10.1002/j.1681-4835.2013.tb00398.x

Bello-Bravo, J., Lutomia, A. N., Madela, L. M., & Pittendrigh, B. R. (2017). Malaria prevention and treatment using educational animations: A case study in Kakamega County, Kenya. *International Journal of Education and Development using Information and Communication Technology, 13*(1), 70–86.

Bello-Bravo, J., Abbott, E. A., Mocumbe, S., & Pittendrigh, B. R. (2020a). Identifying and evaluating farmer deviations from steps recommended for hermetic postharvest storage of beans in northern Mozambique. *Journal of Stored Products Research, 87*, 101628. https://doi.org/10.1016/j.jspr.2020.101628

Bello-Bravo, J., Lutomia, A. N., Abbott, E., Mazur, R., Mocumbe, S., & Pittendrigh, B. R. (2020b). Making agricultural learning accessible: Examining gender in the use of animations via mobile phones. In M. Khosrow-Pour, S. Clarke, M. E. Jennex, & A.-V. Anttiroiko (Eds.), *Environmental and agricultural informatics: Concepts, methodologies, tools, and applications* (pp. 716–736). IGI.

Bello-Bravo, J., Payumo, J., & Pittendrigh, B. R. (2021). Measuring the impact and reach of informal educational videos on YouTube: The case of Scientific Animations Without Borders. *Heliyon, 7*(2021), e08508. https://doi.org/10.1016/j.heliyon.2021.e08508

Bello-Bravo, J., Medendorp, J. W., & Pittendrigh, B. R. (2022). Just participation or just participation? A participatory justice model for more successful theory of change design, implementation, and solution uptake. *Heliyon, 8*(7), e09808. https://doi.org/10.1016/j.heliyon.2022.e09808

Bello-Bravo, J., Medendorp, J. W., Lutomia, A. N., & Pittendrigh, B. R. (2023a). Contents, media, and genres for learning. In *Gender, digitalization, and resilience in international development: Failing forward* (pp. 195–221). Routledge. https://doi.org/10.4324/9781003226383

Bello-Bravo, J., Medendorp, J. W., Lutomia, A. N., & Pittendrigh, B. R. (2023b). Translating to connect local languages for learning. In *Gender, digitalization, and resilience in international development: Failing forward* (pp. 234–254). Routledge. https://doi.org/10.4324/9781003226383

Bello-Bravo, J., Medendorp, J. W., Lutomia, A. N., & Pittendrigh, B. R. (2023c). Wicked solutions for wicked problems: Catalysts and the WIDGET model. In *Gender, digitalization, and resilience in international development: Failing forward* (pp. 129–168). Routledge.

Brewster, F. (2020). *The racial complex: A Jungian perspective on culture and race*. Routledge.

Brock-Utne, B. (2005, December 5–7). The interrelationship between language and culture. In *Evaluation of education—On whose terms?* Beitostølen, Norway.

Byock, J. L. (2001). *Viking Age Iceland*. Penguin UK.

Church-Morel, A., & Bartel-Radic, A. (2016). Skills, identity, and power: The multifaceted concept of language diversity. *Management International, 21*(1), 12–24. https://doi.org/10.7202/1052494ar

Corntassel, J. (2008). Toward sustainable self-determination: Rethinking the contemporary indigenous-rights discourse. *Alternatives, 33*(1), 105–132. https://doi.org/10.1177/030437540803300106

Courington, C. D. (2005). *Postmodernity and learned helplessness: Ideological strategies, political realities* [Thesis, Roosevelt University]. Chicago, IL.

Cullen, A., & Coryn, C. L. (2011). Forms and functions of participatory evaluation in international development: A review of the empirical and theoretical literature. *Journal of Multidisciplinary Evaluation, 7*(16), 32–47. https://doi.org/10.56645/jmde.v7i16.288

Daré, W., Barnaud, C., d'Aquino, P., Étienne, M., Fourage, C., & Souchère, V. (2014). The commodian stance: Interpersonal skills and expertise. In M. Étienne (Ed.), *Companion modelling: A participatory approach to support sustainable development* (pp. 41–67). Springer. https://doi.org/10.1007/978-94-017-8557-0_3

De Vries, G., Terwel, B. W., Ellemers, N., & Daamen, D. D. (2015). Sustainability or profitability? How communicated motives for environmental policy affect public perceptions of corporate greenwashing. *Corporate Social Responsibility and Environmental Management, 22*(3), 142–154. https://doi.org/10.1002/csr.1327

Devi-McGleish, Y., & Cox, D. J. (2018). From weregild to a way forward? English restorative justice in its historical context. *Wolverhampton Law Journal, 1*(1), 21–40.

Dolamo, R. (2013). Botho/Ubuntu: The heart of African ethics. *Scriptura: International Journal of Bible, Religion and Theology in Southern Africa, 112*(1), 1–10. https://doi.org/10.7833/112-0-78

Dragiewicz, M. (2013). Tactics of anti-feminist backlash. In K. Carrington, M. Ball, E. O'Brien, & J. M. Tauri (Eds.), *Crime, justice and social democracy: International perspectives* (pp. 162–174). Palgrave Macmillan.

Dresang, E. T. (2006). Intellectual freedom and libraries: Complexity and change in the twenty-first-century digital environment. *The Library Quarterly, 76*(2), 169–192. https://doi.org/10.1086/506576

Eisler, R. (1994). From domination to partnership: The hidden subtext for sustainable change. *Journal of Organizational Change Management, 7*(4), 32–46. https://doi.org/10.1108/09534819410061360

Ellis, K. F. (1989). *The contested castle: Gothic novels and the subversion of domestic ideology*. University of Illinois Press.

Evans, N., & Levinson, S. C. (2009). The myth of language universals: Language diversity and its importance for cognitive science. *Behavioral and Brain Sciences, 32*(5), 429–448. https://doi.org/10.1017/S0140525X0999094X

Fabbrini, F., & Celeste, E. (2020). The right to be forgotten in the digital age: The challenges of data protection beyond borders. *German Law Journal, 21*(S1), 55–65. https://doi.org/10.1017/glj.2020.14

Fletcher, M. L. (2014). *Tribal justice systems (MSU legal studies research paper)*. Michigan State University.

Freire, P. (1970). *Pedagogy of the oppressed* (M. B. Ramos, Trans.). Continuum.

Gandugade, P. V., Nlooto, M., & Naidoo, P. (2017). Willingness and rationale of traditional health practitioners to collaborate with Allopathic doctors in the eThekwini Metro of Kwa Zulu Natal, South Africa. *PULA: Botswana Journal of African Studies, 31*(1 Supplement 1), 151–161.

Gao, F. (2006). Language is culture: On intercultural communication. *Journal of Language and Linguistics, 5*(1), 58–67. https://doi.org/10.32996/jgcs.2021.1.1.5

Gathogo, J. (2008). African philosophy as expressed in the concepts of hospitality and Ubuntu. *Journal of Theology for Southern Africa, 130*(March), 39–53.

Gawronski, B. (2004). Theory-based bias correction in dispositional inference: The fundamental attribution error is dead, long live the correspondence bias. *European Review of Social Psychology, 15*(1), 183–217. https://doi.org/10.1080/10463280440000026

Gibson, J. (2008). The UDHR and the group: Individual and community rights to culture. *Hamline Journal of Public Law and Policy, 30*(1), 285–313.

Haney, C., & Zimbardo, P. G. (2009). Persistent dispositionalism in interactionist clothing: Fundamental attribution error in explaining prison abuse. *Personality and Social Psychology Bulletin, 35*(6), 807–814. https://doi.org/10.1177/0146167208322864

Heydorn, H.-J. (2024). Survival through Bildung: Outline of a prospect (1974). In H. Kminek (Ed.), *Survival through Bildung: On the topicality of Heinz-Joachim Heydorn's philosophy of education*. Verlag Barbara Budrich.

Humphries, T., Kushalnagar, R., Mathur, G., Napoli, D. J., Padden, C., Rathmann, C., & Smith, S. (2013). The right to language. *The Journal of Law, Medicine & Ethics, 41*(4), 872–884.

Illich, I. (1971). *Deschooling society*. Harper and Row.

Imsen, S. (2013). Law and justice in the realm of the king of Norway. In S. Imsen (Ed.), *Legislation and state formation: Norway and its neighbours in the middle ages* (pp. 15–40). Akademika Publishing.

Jackson, Z. I. (2024). *Becoming human: Matter and meaning in an antiblack world*. New York University Press.

Jefferson, B. (2020). *Digitize and punish: Racial criminalization in the digital age*. University of Minnesota Press.

Johnson, N., Balagamwala, M., Pinkstaff, C., Theis, S., Meinsen-Dick, R., & Agnes, Q. (2018). How do agricultural development projects empower women? Linking strategies with expected outcomes. *Journal of Gender, Agriculture and Food Security (Agri-Gender), 3*, 1–19. https://doi.org/10.22004/ag.econ.293596

Kabeer, N. (1999). Resources, agency, achievements: Reflections on the measurement of women's empowerment. *Development and Change, 30*(3), 435–464. https://doi.org/10.1111/1467-7660.00125

Kelly, F. (1988). *A guide to early Irish law*. Dublin Institute for Advanced Studies.

Kjosavik, D. J. (2021). Sen and development as freedom. In G. Crawford & A.-G. Abdulai (Eds.), *Research handbook on democracy and development* (pp. 79–97). Edward Elgar Publishing.

Knowles, M. S. (1990). *Andragogy in action: Applying modern principles of adult learning*. Jossey-Bass Publishers.

Lee, L. L. (2011). Naat'áanii: What does it mean for Navajo leadership in the 21st century? *Indigenous Policy Journal, XXII*(1), 1–9.

Lerner, G. (1986). *The creation of patriarchy*. Oxford University Press.

Llewellyn, J. J., & Howse, R. (1999). Institutions for restorative justice: The South African truth and reconciliation commission. *The University of Toronto Law Journal, 49*(3), 355–388.

Longman, T. (2009). An assessment of Rwanda's gacaca courts. *Peace Review, 21*(3), 304–312. https://doi.org/10.1080/10402650903099369

Lutomia, A. N. (2019). *A case study of successes and challenges in a scientific collaboration program based in the United States and Benin* [Dissertation, University of Illinois Urbana-Champaign]. Urbana, IL.

Lutomia, A. N., & Bello-Bravo, J. (2024). Increasing lifelong learning using video animations: The case of an agriculture WhatsApp Group in Kenya. In P. Filippo Gomez (Ed.), *Lifelong learning: Education for the future world* (pp. 1–18). IntechOpen. https://doi.org/10.5772/intechopen.114379

Lutomia, A. N., Sibeyo, D., & Lutomia, N. I. (2018). Bulala as an Ubuntu-inspired approach to enhancing organizational culture in rural Kenya. *Journal of Pan African Studies, 11*(4), 102–121.

Lutomia, A. N., Bello-Bravo, J., Medendorp, J. W., & Pittendrigh, B. R. (2020). A positive project outcome: Lessons from a non-dominant government university-based program. *Interdisciplinary Journal of Partnership Studies, 7*(2), 3. https://doi.org/10.24926/ijps.v7i2.3482

Lutomia, A. N., Bello-Bravo, J., Lutomia, N. I., Kataru, J. K., Medendorp, J. W., & Pittendrigh, B. R. (2024). Participatory practices and lessons from Scientific Animations Without Borders and a WhatsApp network in a post-COVID age: The case of video animations for rural agriculture. In A. Chowdhury & G. A. Gow (Eds.), *Digital communication for agricultural and rural development* (pp. 253–270). Routledge.

Malesevic, S. (2013). *Nation-states and nationalisms: Organization, ideology and solidarity*. Polity.

Mangena, F. (2013). Discerning moral status in the African environment. *Phronimon, 14*(2), 25–44.

McDonald, L. (1998). Can collective and individual rights coexist? *Melbourne University Law Review, 22*, 310–336.

Medendorp, J. W., Reeves, N. P., Sal y Rosas Celi, V. G., Harun-ar-Rashid, M., Krupnik, T. J., Lutomia, A. N., Pittendrigh, B., & Bello-Bravo, J. (2022). Large-scale rollout of information and communication technology-enhanced extension training in Bangladesh demonstrates challenges and opportunities towards inclusive gender participation. *PLoS One, 17*(7), e0270662. https://doi.org/10.1371/journal.pone.0270662

Medendorp, J. W., Bello-Bravo, J., Lutomia, A. N., & Pittendrigh, B. R. (2023). The medium for the missing: Participatory animated media for ethical and effective information dissemination. In J. Bello-Bravo, J. W. Medendorp, A. N. Lutomia, & B. R. Pittendrigh (Eds.), *Gender, digitalization, and resilience in international development: Failing forward* (pp. 80–102). Routledge.

Memarian, B., & Doleck, T. (2023). Fairness, accountability, transparency, and ethics (FATE) in artificial intelligence (AI), and higher education: A systematic review. *Computers and Education: Artificial Intelligence, 5*, 100152. https://doi.org/10.1016/j.caeai.2023.100152

Mezirow, J. (1991). *Transformative dimensions of adult education*. Jossey-Bass.

Mills, A., Drake, K., & Muthusamipillai, T. (2017). An Anishinaabe constitutional order. In P. Smith (Ed.), *Reconciliation in Canadian courts: A guide for judges to aboriginal and indigenous law, context and practice* (pp. 1–18). National Judicial Institute.

Morgensen, S. L. (2011). The biopolitics of settler colonialism: Right here, right now. *Settler Colonial Studies, 1*(1), 52–76. https://doi.org/10.1080/2201473X.2011.10648801

Nemutandani, S. M., Hendricks, S. J., & Mulaudzi, M. F. (2016). Perceptions and experiences of allopathic health practitioners on collaboration with traditional health practitioners in post-apartheid South Africa. *African Journal of Primary Health Care & Family Medicine, 8*(2), 1–8.

Nonaka, I., & Takeuchi, H. (1997). *The knowledge-creating company: How Japanese companies create the dynamics of innovation*. Oxford University Press.

NSN. (2024). *Hózhǫ́ji naat'aah (Diné Traditional Peacemaking)*. https://courts.navajo-nsn.gov/Peacemaking/Plan/peace.html

Ortiz, A. A., Fránquiz, M. E., & Lara, G. P. (2020). Co-editors' introduction: Culture is language and language is culture. *Journal of the National Association for Bilingual Education, 43*(1), 1–5. https://doi.org/10.1080/15235882.2020.1741303

Ovando, C. (1990). Essay reviews: Politics and pedagogy: The case of bilingual education. *Harvard Educational Review, 60*(3), 341–357.

Pálsson, H., & Edwards, P. (1972). *The book of settlements*. University of Manitoba Press.

Payumo, J., Bello-Bravo, J., & Pittendrigh, B. R. (2021). Demonstrating the nexus effects of online videos, research outputs, and investments to knowledge absorption using linguistically adapted animations. *European Journal of Applied Sciences, 9*(2), 283–301. https://doi.org/10.14738/aivp.92.10025

Paz, M. (2013). The failed promise of language rights: A critique of the international language rights regime. *Harvard International Law Journal, 54*(1), 157–218.

Peabody Awards. (1954). *Man's right to knowledge*. https://peabodyawards.com/award-profile/mans-right-to-knowledge/

Peters, M. A., Lawton, F. D., Kapitzke, C., Gietzen, G., & Willinsky, J. (2007). Review symposium: The access principle: The case for open access to research and scholarship. *Policy Futures in Education, 5*(3), 401–423. https://doi.org/10.2304/pfie.2007.5.3.401

Piatt, B. (1986). Toward domestic recognition of a human right to language. *Houston Law Review, 23*, 885–906.

Pjevovic, D. (2017). *Cities and climate change: Power games and greenwashing through transnational urban networks* [University of British Columbia]. Vancouver, Canada.

Popowich, S. (2020). The problem of neutrality and intellectual freedom: The case of libraries. In C. L. Riley (Ed.), *The free speech wars: How did we get here and why does it matter?* (pp. 43–52). Manchester University Press. https://doi.org/10.7765/9781526152558.00008

Rawls, J. (1971). *A theory of justice*. Harvard University Press.

Reeves, H., Grandjean, L., Razafindrina, K., Amiadanarivo, A., Rakotosamimanana, N., Razafindranaivo, T., Seimon, T., Blalock, B., Bello-Bravo, J., Pittendrigh, B. R., & Small, P. (2020). Evaluating the use of educational videos to support the tuberculosis care cascade in remote Madagascar. *International Journal of Tuberculosis and Lung Disease, 24*(1), 28–35. https://doi.org/10.5588/ijtld.19.0161

Reynolds, I. (2021). *Ideology, critique, and political education* [Dissertation, Villanova University]. Villanova, PA.

Robertson, D. L. (2015). Invisibility in the color-blind era: Examining legitimized racism against indigenous peoples. *American Indian Quarterly, 39*(2), 113–153.

Roediger, D. R. (2019). *How race survived US history: From settlement and slavery to the Obama phenomenon*. Verso.

Sanya, B. N. (2017). *States of discretion: Black migrating bodies and citizenship in the United States* [Ph.D., University of Illinois at Urbana-Champaign]. Champaign, IL.

Sen, A. (2012). Values and justice. *Journal of Economic Methodology, 19*(2), 101–108. https://doi.org/10.1080/1350178X.2012.683601

Sillitoe, P. (2002). Globalizing indigenous knowledge. In P. Sillitoe, A. Bicker, & J. Pottier (Eds.), *Participating in development: Approaches to indigenous knowledge* (pp. 108–138). Routledge.

Skiba, R. J., Arredondo, M. I., & Williams, N. T. (2014). More than a metaphor: The contribution of exclusionary discipline to a school-to-prison pipeline. *Equity & Excellence in Education, 47*(4), 546–564. https://doi.org/10.1080/10665684.2014.958965

Souto-Otero, M. (2021). Validation of non-formal and informal learning in formal education: Covert and overt. *European Journal of Education, 56*(3), 365–379. https://doi.org/10.1111/ejed.12464

Stanhope, G. W. (2024). *Justice in the Icelandic commonwealth* [Thesis, Eastern Kentucky University]. Richmond, KY.

Tauri, J., & Morris, A. (1997). Re-forming justice: The potential of Maori processes. *Australian & New Zealand Journal of Criminology, 30*(2), 149–167. https://doi.org/10.1177/000486589703000203

Tudor, S. L. (2013). Formal–non-formal–informal in education. *Procedia-Social and Behavioral Sciences, 76*, 821–826. https://doi.org/10.1016/j.sbspro.2013.04.213

UNESCO. (2010). *Freedom of information: The right to know*. UNESCO.

UNESCO. (2022). *International decade of indigenous languages 2022–2032*. https://www.un.org/development/desa/indigenouspeoples/indigenous-languages.html

Van Wormer, K. (2008). Anti-feminist backlash and violence against women worldwide. *Social Work & Society, 6*(2), 324–337.

Wagner, S. R. (2011). *Sisters in spirit: Haudenosaunee (Iroquois) influence on early American feminists*. Native Voices Books.

Waller, D. M., & Reo, N. J. (2018). First stewards: Ecological outcomes of forest and wildlife stewardship by indigenous peoples of Wisconsin, USA. *Ecology and Society, 23*(1), 45. https://doi.org/10.5751/ES-09865-230145

Watson, B. (2017). The troubling evolution of corporate greenwashing. *Chain Reaction, 129*, 38–40. https://doi.org/10.3316/ielapa.766428450523476

Williams, R. (1980). *Problems in materialism and culture: Selected essays*. Verso.

Williamson, S. M., & Prybutok, V. (2024). Balancing privacy and progress: A review of privacy challenges, systemic oversight, and patient perceptions in AI-driven healthcare. *Applied Sciences, 14*(2), 675. https://doi.org/10.3390/app14020675

Yeung, A. W., Kay, A. C., & Peach, J. M. (2014). Anti-feminist backlash: The role of system justification in the rejection of feminism. *Group Processes & Intergroup Relations, 17*(4), 474–484. https://doi.org/10.1177/1368430213514121

Yosso, T. J. (2005). Whose culture has capital? A critical race theory discussion of community cultural wealth. *Race Ethnicity and Education, 8*(1), 69–91. https://doi.org/10.1080/1361332052000341006

Zehr, H. (1990). *Changing lenses: A new focus for crime and justice*. Herald Press.

Chapter 7
Conclusion: Preservation, Sustainability, the Right to Knowledge, and the SDGs

Introduction

Sustainability requires that our generation's enjoyment of the world's goods and gifts does not compromise future generations' ability to enjoy them as well. The concept of sustainability, then, represents a principle of intergenerational equity—along with giving overriding priority to meeting the needs of the world's poorer and recognizing the limits of technology (Brundtland et al., 1987). As an approach, a goal, and a core aspect of the Sustainable Development Goals (SDGs), sustainability applies to the entire spectrum of environmental, cultural, and intellectual goods and gifts of the world, underscoring a collective responsibility to steward these resources wisely.

Recognizing our role as caretakers rather than consumers at all scales—from the general public to global organizations and researchers—we are called to navigate how best to meet both our own needs and those of future inhabitants of the planet, including other species. This approach promotes the conservation not only of physical resources but also of cultural and social heritage, as emphasized in triple-bottom-line decision-making frameworks (Longoni & Cagliano, 2018; Svensson et al., 2018). Such decision-making aligns with the medical principle of doing no harm, aiming for interventions that are maximally effective and minimally invasive in the short term, while having the fewest and smallest long-term detrimental effects. Ethical sustainability, in this sense, works to ensure that today's actions do not undermine tomorrow's life possibilities, opportunities, or well-being.

For the right to knowledge, this ideal means not only preserving currently available information so it is not lost but also ensuring it remains conveniently usable in the future. While some of these archives will also exist physically and in the memories of living people, digital preservation and access—through future iterations of the Internet of Things (IoT) and vast networks of hardware and software comprising

J. Bello-Bravo et al., *The Right to Knowledge*,
https://doi.org/10.1007/978-3-032-03360-4_7

the global digital infrastructure—will undoubtedly become the most prevalent form, provided sustainable sources of electricity remain.

Already, the library sciences provide substantial expertise for curating any archives of gold-standard, life-sustaining knowledge such that future patrons of that library can use them for solution-seeking when and where they are wanted. Technical challenges—including the full digitization of material and immaterial knowledge, interfaces for accessing it (virtual or otherwise), and especially the linguistic translation of knowledge into forms anyone can use—are increasingly solvable, even as their automation remains much in need of improvement (especially in cases where exact translations are critical). More difficult will be the ongoing stewardship of the archive: incorporating new entries, updating existing ones, eliminating obsolete material, making governance decisions around these processes, and ensuring continued accessibility—especially to search algorithms—as digital technologies, including AI, continue to evolve.

In this case, the imperative to give overriding priority to meeting the needs of the world's poor underscores why a sustainable archive for the future must prioritize the preservation of indigenous knowledge and practices—not only to ensure these are not lost and remain usable by future generations, but also because such traditions have sustained human societies for orders of magnitude longer than current systems of industrial knowledge. This requires acknowledging *who* is empowered to produce knowledge. Without endorsing the full program of Foucault's power/knowledge analysis, we must concede that there has historically been an inevitable relationship between who holds power and who defines knowledge (Leps, 1992).

This tension is reflected in every textbook in every classroom in every school in the world. History is written by the victors, and regimes determine what is important to learn and what "knowledge" will be produced and shared to shape the next generation. As Locke noted in his *Two Treatises of Government* (1690), knowledge is the servant of the regime, but it can also be a lever of resistance and a defense of human rights. Efforts to preserve historical "knowledge" are not only part of our heritage but also essential to our future.

As an example, in the 1930s, the father-and-son team of John and Alan Lomax traveled through the southern United States to document and preserve folk music traditions. Many of their recordings were made in southern prisons. Listening to these archives today, one hears voices that might otherwise have been lost. The individuals recorded often lived without social power or recognition, yet their songs speak vividly of lived experiences that would have remained widely unheard except for the commitment to their preservation and, most importantly, the generosity and courage of those who chose to share their voices.

These archives now help us understand the roots of virtually every form of contemporary music in the United States (Stone, 2021). They speak to deep spirituality, reflections on suffering and injustice, and a questioning of the way the world is and why it remains that way. Within them are embedded philosophies that prompt us to reconsider assumptions about what constitutes valid knowledge and invite us to engage with the lived realities of people denied recognition in the validated knowledge of official histories—yet whose influences are present in every aspect of US culture (Stone, 2021).

In the historical relationship between power and knowledge, important information was often lost simply because it was not recognized as "knowledge." With the rise of the IoT, however, this need no longer be the case (Dachyar et al., 2019). While vested interests still attempt to control or devalue certain forms of knowledge, doing so may be becoming more untenable, if not increasingly undesirable. Time will tell. In the meantime, social media platforms like TikTok, often dismissed with an eye-roll, are saturated with human experience and offer a rich field for psychological and sociological insight (Li et al., 2024; Sattora et al., 2024). Preserving such everyday chronicles may be essential for understanding ourselves and shaping a more inclusive collective future. But such preservation will not happen automatically. It will require deliberate and sustained effort to ensure that the experiences and viewpoints are curated, analyzable, and accurately reflect those experiences. Learning from the past to shape the future requires committing to safeguarding the perspectives most at risk of being erased, whether through suppression, neglect, or simple disregard.

More critically, none of this will matter if the servers go dark. Knowledge incurs a cost, and with the rise of AI, that cost, in terms of energy alone, is increasing daily. The Information Age's industrial dependence on finite fossil fuels for electricity—and its resultant climate damage—means that expending those resources now robs future generations of access later (Hensley & Steer, 2019). Although often treated as a truism, the reality remains: sustainable knowledge—keeping archives online, powering data centers, ensuring global access—requires a new energy paradigm now, not later. The knowledge of the current future will not survive on the energy systems of its present past.

The Candle of Knowledge

The transfer of knowledge—especially when loosely coupled with digital ICTs—can lead to unprecedented opportunities for innovation and problem-solving. As a renewable and expandable resource, sustainably archived and curated knowledge provides a foundation for attaining the SDGs and enables communities and societies to resiliently adapt to emergent challenges. Stewardship of life-sustaining knowledge thus requires a commitment not only to its preservation and active dissemination but also to its open production and indefinite usability through people's exercise of their right to it.

These goals critically intersect with the materiality of knowledge. Though often imagined in abstract terms, knowledge is always embodied (Oyama, 2000)—whether in "inert" forms (e.g., print, animated videos, digital archives, sculptures, buildings, musical scores and recordings, films, and scripts) or in "living" forms (e.g., human memory, embodied practice, rituals, and performances). A central strength and limitation of inert knowledge is its permanence: what a book or file says today is what it will say tomorrow (assuming no media degradation), even if that content has been superseded by better knowledge. Conversely, a central strength and limitation of living knowledge lies in its adaptability—as skills and practices that evolve to meet new contexts but in doing so risk forgetting or overwriting what was once known. The further tension then between inert and living knowledge is a need for careful discernment: when should we hold the line of knowledge and when must we allow transformation?

In *Orality and Literacy*, Water Ong (1982) provides a striking example of this tension. In a British colony in Africa, two tribal factions brought a land dispute before a British magistrate. The magistrate noted that neither faction's memory nor their accounts of the past aligned with the written records compiled by British colonizers. So, whose knowledge prevailed? Ong points out that the magistrate's reliance on colonial documentation failed to account for the lived, oral reality of "history" as it existed between the parties. The dispute, in short, did not hinge on some inert, unchanging past, but about resolving and evolving a living narrative to settle a present-day conflict. This becomes urgent in cases where indigenous peoples, whose languages may have no written form, are forced to present their claims before legal or bureaucratic systems that recognize only written documentation as legitimate knowledge. Such privileging allows indigenous knowledge and practice—despite generations of empirically validated and effective use—to be dismissed or delayed, pending scientific validation.

In this sense, knowledge—specifically, life-sustaining knowledge—is not a neutral "resource" but an argument, a warrant for doing or not doing something. Recognizing this dimension is essential for sustaining the right to knowledge. Future generations will only inherit what we choose to document today, but that documentation must also reflect and be relevant to the living, contested, and contextual facets of what is recognized as "knowledge."

Notwithstanding this, it is a remarkable fact that knowledge can be shared without depleting it. The metaphor of a lit candle lighting another often symbolizes this potential of knowledge to grow and spread without diminishment. The image captures the essence of how knowledge can be passed on, expanding its reach, engagement, and empowerment while remaining intact. Unlike finite resources, the act of sharing knowledge can lead to an amplification of understanding and innovation, as each new knowledge recipient then carries the potential to adopt—if not also build upon and adapt—what they have learned toward reaching core SDGs: eliminating poverty, hunger, and disease, and sustainably increasing food and gender security and community well-being.

The process of knowledge transfer is readily more than additive, i.e., multiplicative (or even exponential), when networks of informed people are enabled to address complex issues collaboratively and become sharers and teachers of life-sustaining knowledge themselves. Thus, a sustainable stewardship of life-sustaining knowledge involves creating systems and structures to facilitate its unhindered access and application, enabling people and communities to benefit from a cumulative and ever-growing body of wisdom through a practical application of knowledge. This is the exercise of a right to knowledge at its core.

However, it is necessary to look more closely at this common metaphor of lit candles passing their flames from one to another. As with metaphors generally (Lakoff & Johnson, 1980), their clarity can conceal as much as they reveal. For example, we might ask: who possesses these candles for sharing in the first place? Who makes them? Who distributes them? Posing the question already interrupts the metaphor's smooth surface and exposes historical and structural inequalities in access to "candles," e.g., education, resources, and the institutions that regulate and certify knowledge. The metaphor, in its idealized form, presumes a world where such access is universal, where the act of lighting one candle with another is both effortless and universally available. But this façade is precisely what is not the case.

The metaphor also isolates the moment of ignition—the touching of one wick to another—as if this is the whole story. For that flame to catch, a whole set of conditions must be met: proximity, intention, preparation, and a stable environment in which the exchange can take place. These factors are not incidental but constitutive and underscore the central importance of lowering transactional costs to make such proximity, intention, preparation, and environmental stability possible. Knowledge does not transfer without labor and effort, and consent, willingness, and readiness on the part of the learner. This work recalls what adult learning theorists have long affirmed: that meaningful learning depends not simply on access, but on participation—voluntary, situated, and contextual (Bangura, 2005; Knowles, 1990; Ozuah, 2016).

What the image of the candle elides are the conditions that make the moment of transmission possible—the socioeconomic and institutional infrastructures that make the act of learning possible in the first place. It presents the learning effected by knowledge transfer as something free-floating, ready to be passed along in any context, at any time, by anyone. History and experience suggest otherwise; knowledge does not circulate evenly, equitably, or freely, especially not to women whose hindered participation in or exclusion from education depends on a web of factors not visible in the metaphor. One might even ask: whose hand holds those candles? Why don't some wicks catch? What happens when gusts of wind blow both out? How do we avoid that?

One must also interrogate the central presumption of the metaphor: that knowledge is, in fact, *transferred*. This seemingly minor or overly pedantic objection asks whether the flame on the second wick is factually the same as the flame on the first. Following the constructivist insights on learning (Ashby, 1956; von Glasersfeld, 1991; Vygotsky, 1978) and critiques of the "banking" model of education (Freire, 1970; Gadotti, 2017), the answer cannot be *yes*. Concretely, we see this illustrated in "deviations" of Mozambican farmers using the jerrycan method, where 20% skipped at least one protocol step and 33% added a step, without the method failing (Bello-Bravo et al., 2020). This example perfectly illustrates the difference between *adapting* an innovation rather than *adopting* it, and how knowledge is not a flame's substance or even its process passed intact from the wick of one mind to another.

The premise of this book is that the exercise of the human right to knowledge by digital communities of practice—loosely coupled with learning systems that are themselves adaptive to change—offers the situational infrastructure necessary to achieve the kind of multiplicative, even exponential, scaling of knowledge dissemination evoked by the candle metaphor. Yet the flickering of that candle's flame also serves as a reminder of the fragility and vulnerability of the institutions and practices required to sustain such growth—especially now, in a moment when the field of international development is being radically reshaped. Just as candles require tending, trimming, and, eventually, replacement to continue burning, so too does the production of knowledge demand care, stewardship, and the capacity for organizational transformation—even, at times, a willingness to give up existing structures altogether.

This challenge involves not only the preservation of intellectual practices—indigenous, scientific, and otherwise—but also the active cultivation of values that make knowledge vital: lifelong learning, critical reflection, and a curiosity oriented toward the new (Lutomia & Bello-Bravo, 2024). At the heart of this interest is a seldom-emphasized truth: the desire for a better world necessarily also requires a willingness to relinquish the precarious comforts and settled routines of the current one.

Changing habits is rarely simple, even for people who genuinely wish to. But there are also powerful forces at play—economic, institutional, ideological—that depend on the continuity of those habits. For them, the status quo is not an obstacle but the very condition of profitability and survival. While present upheavals in international development may force some degree of reckoning and change, hopes for transformation usually tend to be projected into the future. It is often imagined that the next generations, not us, will take the necessary steps—that change will arrive not through disruption *now*, but through a slow cultivation of new desires and values *then*.

If our affection and concern for those next generations—frustrating as they may sometimes be currently—can motivate us to meet sustainability's challenge and safeguard their access to life's goods, then we must also be wary of turning our concern into an excuse. Hope for the future must not become justification for delay. To put it plainly: we cannot keep kicking the can down the road (as it was kicked to us) and keep continuing to call that sustainability.

Instead, by establishing times and places now—both physical and digital—where life-sustaining knowledge can be shared, discussed, learned, elaborated, and improved is essential for keeping its candle flame lit. While our hope is still that this flame could refer to knowledge in its broadest sense, at present, we insist on knowledge in its most basic sense: as gold-standard, life-sustaining information and practices to support the core of the SDGs' human dignity: no hunger, good health and well-being, safe shelter, gender security, and access to knowledge that ensures no one experiences an impoverished life—as knowledge that enables the world's poorer to exercise their right to such knowledge unhindered by the self-interests of the less poor. In this sense, a sustainability of knowledge is not just about the information itself but also about the communities, cultures, and learning systems that ensure its social reproduction.

The Demand Side of Sustainability

Pursuing the SDGs and the foundational right to knowledge cannot be accomplished by archiving alone. The sustainability of knowledge necessarily involves decision-making around its use now with respect to the future (Gibson, 2006). A classically *pragmatist* definition of knowledge is that it is a true, justified belief, but that truth and justification must be rooted in a predictive power when applied—or more briefly, that it works. How do I know, for example, that 2 + 2 = 4? Not simply because it is written in a book, but because when someone places two apples next to two more, I reliably observe the outcome: there are four apples.

This deliberately oversimplified example nonetheless points to the more broadly established school of thought represented by pragmatist thinkers ranging from John Dewey, Charles Pierce, and William James to contemporary theorists like Cheryl Misak, Richard Rorty, Hilary Putnam, and Huw Price (Misak, 2007; Thayer, 1982). On this view, knowledge must serve a functional role in prediction, explanation, or control of our reality. Empirical knowledge broadly is confirmed through repeated application and practical success, while scientific knowledge, more rigorously, is tested through experimentation, falsifiability, and peer review. Without these empirical processes around what works, we are left only with conjecture—and sometimes that is all we can manage. Ideally, pragmatic knowledge enters our shared lexicon when its claims hold up in use, when prediction becomes confirmation. Without such application, even the most carefully preserved ideas risk fading into irrelevance.

It is crucial to recognize that grounding knowledge in pragmatic criteria, i.e., in what works in practice, gives no ground or license to dismiss or marginalize indigenous forms of knowledge and practice. On the contrary, many indigenous knowledge systems have demonstrated their reliability and effectiveness over centuries, often through precisely the kind of situated, trial-based adaptation that pragmatist thinkers value. To measure the validity of indigenous knowledge solely by external, technoscientific standards would be not only epistemologically narrow but also ethically suspect.

At the same time, it must be acknowledged that pragmatic validation is neither always straightforward nor universally applicable. Some forms of moral, aesthetic, spiritual, and cultural knowledge resist predictive framing altogether, and forcing them into that mold risks erasing the very meanings that they sustain. An honest commitment to the right to knowledge requires both respect for the practical efficacy of diverse epistemologies and humility about the limits of pragmatic verification. In the context of the SDGs, these parameters mean focusing on knowledge in its life-sustaining forms—knowledge embedded in practice, community, and survival—not abstract formulations alone.

More exactly, it is overly tendentious to starkly draw a line between abstract and applied knowledge. All applied knowledge will be abstractly described in practice through some cultural lens, and every cultural abstraction can be tested for what works or not in applications. The placebo effect offers an excellent example where one is well-advised to use medicines and techniques that inexplicably work despite any adequately empirical reason why (Horwitz et al., 2017). At the same time, we recognize that today's abstractions may find application tomorrow. Our emphasis on pragmatism is not a rejection of the purely theoretical but a reminder to prioritize research that responds creatively and concretely to meeting the needs of the world's poor and the SDGs. Sustainability, in this context, means knowledge that is convenient, relevant, and usable *now*, not in some arguably speculative future. This demand is not a dismissal of intellectual exploration but a reminder that knowledge must also meet the life-sustaining urgency of its moment.

Indeed, we must locate the primary validity of life-sustaining knowledge to fulfill the SDGs on the demand side. Building and maintaining a repository of knowledge available to posterity is necessary but not sufficient. Rather, it requires doing what we must to allocate the methods at our disposal toward ensuring that knowledge is deliverable at the times and in the places for anyone but especially those people who need it most. Only when such an encyclopedia or repository is on a "shelf" from which it can be taken down, perused, and used by anyone will it thoroughly contribute to people's exercise of their right to life-sustaining knowledge and the achievement of the SDGs.

Knowledge Societies and the Right to Knowledge

From the time that thinkers began to outline the characteristics of a post-industrial transition to an Information Age, speculation on emergent information societies (Bell, 1973; Castells, 1997; Machlup, 1972; Toffler, 1991; Touraine, 1971; Umesao, 1963), learning societies (Delors, 1998; Faure, 1972; UNESCO, 2005), knowledge societies (Petrushenko & Vorontsova, 2020), and their attendant knowledge management systems (Drucker, 1969; Leonard, 1995; Rogers, 1986; Strassman, 1994) fundamentally re-framed the view of society in five key ways—ways that did not merely support but facilitate meeting the SDGs.

First, the foundational shift from an industrial to post-industrial society views development as no longer primarily about material infrastructure but about information flows, capacity to learn, and the equitable distribution of knowledge resources. This re-vision directly supports SDG 9 (Industry, Innovation and Infrastructure) and SDG 8 (Decent Work and Economic Growth), but now through the vector of digital, service-based, and knowledge-driven economies. Second, the emphasis on access—to ICTs, to educational opportunities, to the tools of knowledge production—aligns with SDG 4 (Quality Education) and SDG 10 (Reduced Inequalities). But this is not just about formal schooling. The transition from an "information society" to a "knowledge society" marks a shift from the provision of data to the empowerment of people to use that data meaningfully. That switch directly supports SDG 16 (Peace, Justice, and Strong Institutions) by grounding human rights, especially freedom of expression, access to information, as developmental levers.

Third, the right to knowledge, though not always explicitly named, is embedded as a necessary condition for achieving the SDGs as a whole. The knowledge management loop—generation, storage, distribution, and application—parallels the logic of SDG 17 (Partnerships for the Goals). It offers an operational backbone for cross-sectoral collaboration, collective intelligence, and innovation ecosystems. Fourth, by integrating cultural and linguistic diversity, the framework upholds SDG 11 (Sustainable Cities and Communities) and SDG 5 (Gender Equality), because inclusion in knowledge societies cannot be reduced to technical access alone. Who is allowed to generate knowledge? Whose voices are heard? This framework addresses the conditions of possibility for participation itself, pointing toward structural transformation, not just policy intervention. Lastly, knowledge management supplies methods, feedback loops, and learning systems necessary to track progress, adapt strategies, and avoid stagnation. In this way, the knowledge society represents not just a single SDG initiative among others but meta-architecture through which all of the goals become legible, actionable, and responsive.

While these knowledge society principles align well with SAWBO's objectives and practices overall, SAWBO's learning-systems approach (described in Chap. 2) makes the flow of information production and reception *bidirectional*. The individual elements that comprise this bidirectional learning-systems approach are not unique to SAWBO; rather, their combination and systems-configuration are. Nor does SAWBO alone have this capability; a goal of this book is to enable others to adapt its learning-systems approach to help implement the technical capacity for attaining the SDGs and to empower people to exercise their right to knowledge and improve their and their communities' well-being and quality of life everywhere. However, SAWBO's integrated systems approach focuses on actualizing the right knowledge for those who speak local languages, may not be literate, or may exist on the "wrong side" of the digital divide (Fig. 7.1).

Life-Sustaining Knowledge and the SDGs

For knowledge to fulfill its role as a life-sustaining right, it must be treated not as an abstract ideal or a cache of inaccessible data, but as a pragmatic social good, e.g., a universally available digital encyclopedia, ready to be pulled from the shelf by anyone, anywhere, at any time. However, this knowledge must be prioritized according to human needs, with basic and foundational needs addressed first. Failing to do so risks not just stalling progress on the SDGs but actively undermining them. People cannot access education (SDG 4), innovation (SDG 9), or justice (SDG 16) when they are starving (SDG 2), sick (SDG 3), or unsafe, for example, due to a lack of clean water (SDG 6), electricity (SDG 7), or gender security (SDG 5).

Various scholars have mapped this prioritization of hierarchically arranged needs in terms of basic ones (e.g., food, water, shelter, and clothing), physical and emotional safety, emotional belonging and well-being, self and social esteem, and self-actualization (Nikolova, 2024; Panagiotopoulos et al., 2024; Tian et al., 2024). Provided that such framings center human dignity and are social—not overly individualistic (McLeod, 2007), consistent with human rights discourse, as a *basic* need

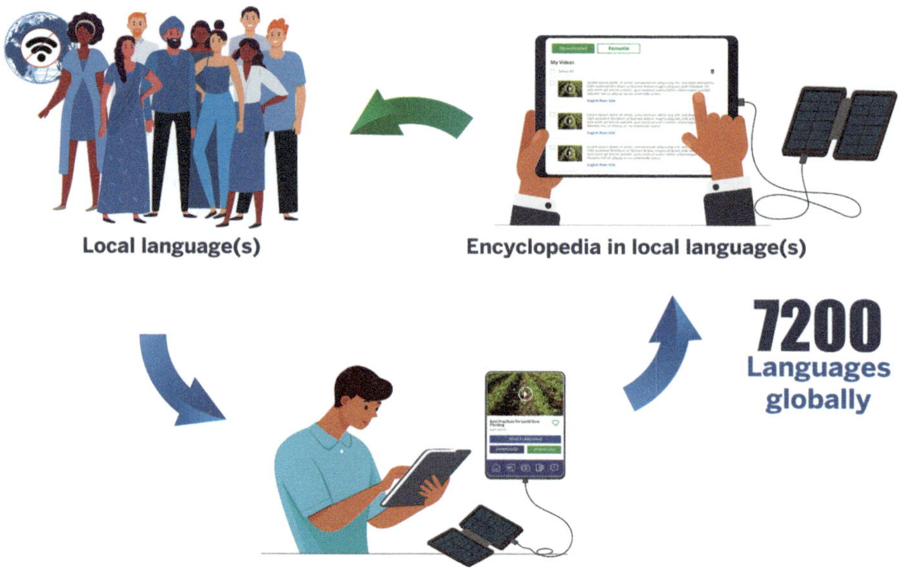

Fig. 7.1 Integrated mobile app, local devices, and solar chargers system. SAWBO's mobile phone app offers an integrated learning-systems approach that enables organizations and change agents to deliver video libraries—translated into the upper bound of potentially 7200 world languages—to communities without access to smartphones, electricity, or the Internet. By preloading the app onto a tablet and downloading videos in relevant languages, it creates a portable library of gold-standard R4D knowledge for local use. A solar-powered charger allows the tablet to function in off-grid areas, and after a short training session, the device can be left with the community. Once the content has been used, the tablet and charger can be returned to the parent organization for updating and redeployment to other communities. Reprinted with permission from SAWBO (2025)

(de Oliveira Cintra Farias & Ferreira, 2024)—then this hierarchy helps sustainability decision-making around what needs must be first met by knowledge, providing a foundation for meeting other needs. To date, the most accessed SAWBO videos for which data is available include "Survival Gardening: Drip Irrigation" (SDG 2), "Charcoal Water Filtration" (SDG 2), "Tuberculosis Prevention" (SDG 3) (Rodríguez-Domenech et al., 2023), and the use of hermetically sealed jerrycans to prevent postharvest losses and increase food security (SDG 2) (Bello-Bravo et al., 2022)—basic needs and physical well-being.

The Future of the Right to Sustainable Knowledge

To uphold the right to sustainable knowledge for future generations, it is crucial to create systems that preserve and enhance knowledge in ways that are resilient and adaptable but also accessible and useable. Preserving and ensuring access to existing information while sustainably fostering the development of new knowledge (Bello-Bravo, 2023) delivers that knowledge to those who most need it in scalable, cost-effective ways. While such systems must be sturdy enough to withstand the test of time and the challenges posed by changing social, economic, and environmental conditions, they must also be "willing" to cease to exist organizationally in their current form so as to re-form in new, loosely coupled systems with others.

In its most self-evident form, a sustainable system must be non-exclusive—built on the premise that everyone, regardless of background or circumstance, can exercise their right to knowledge and participate in, as well as benefit from, the prevailing body of knowledge. We stress again the pragmatic emphasis here: the most universally accessible knowledge must be life-sustaining knowledge. We will not accept, on any grounds, that people can be rationally denied access to knowledge that enables better food, health, shelter, gender security, or well-being. In the present era, this insistence requires prioritizing open access and guaranteeing that digital literacy is not a prohibitive barrier; in future eras, this specific pragmatic emphasis may change, but the underlying imperative will not.

Ultimately, the objective of cultivating a culture of sustainability—both generally and more narrowly around knowledge production and dissemination specifically—is an impossible bootstrapping if it does not include the prospect of transforming into something currently unforeseeable, and perhaps even disagreeable to what we currently consider desirable. This point is simply to observe that what genuinely constitutes a more sustainable world in the future may involve individual and institutional/corporate forms of personhood, behavior, and sustainability not currently valorized. The learning-systems approach we advance pragmatically in this book recognizes and operationalizes movement toward such a transformed goal.

Conclusion

Achieving the SDGs pragmatically demands an emphatic focus on knowledge that not only *works* but also centers on foundational needs—those that, when fulfilled, support life, strengthen communities, sustain ecosystems, and provide a basis for meeting more complex, downstream challenges. In all cases, these criteria require value chains that lower the transactional costs of moving gold-standard basic research off the shelf, through solution development, deployment, and adaptation by end-recipients when, where, and in whatever way everyone can use. This need is more urgent than ever, given the sector-wide impacts affecting international development into the foreseeable future.

Prioritization of *life-sustaining knowledge that works* shifts the focus away from any urgency about resolving an indigenous versus scientific knowledge debate (Agrawal, 2002) and instead toward de-enabling when "scientific" knowledge deprecates or refuses to acknowledge "indigenous" knowledge that empirically works (Horwitz et al., 2017; Nemutandani et al., 2016). Here, it is not only that indigenous practices and knowledge production—adapted through long-term responses to local environments and settler-colonial impacts (Trosper, 2007)—offer high-priority, sustainability-aligned methodologies and solutions (Golo & Yaro, 2013; Mathevet et al., 2018; Messier et al., 2015) but also decision-making schema for policymaking around forest stewardship (SDG 15) or national energy (SDG 7) (Kimmerer, 2012; Waller & Reo, 2018).

Often premised in notions of balance, reciprocity, and a recognition of the "moral status" of not-only-human life in the world (SDG 14 and SDG 15) (Mangena, 2013), these indigenous notions readily harmonize with sustainability's premises and the ethos of the SDGs (Adeyeye, 2019; Kumar et al., 2021; Sandoval-Rivera, 2020). Achieving their outcomes does not require wholesale adoption of these perspectives (Bello-Bravo, 2020), just a refusal to dismiss them for being different (which is inhumane) or to ignore where they work (which is irrational, especially under the technocratic premise to scientifically optimize the best solutions to problems).

The premise of this book is that the fulfillment of the SDGs will remain elusive if people cannot exercise their right to life-sustaining knowledge. We have described at length how the solution to the technical aspect of this problem is already solved and being implemented around the globe, while indicating its still-unsolved or "wicked" social aspect. Further advances in research to expand the existing library and more widely mass-scale existing solutions will extend its growing global reach. But its progress will be needlessly slowed so long as the problem's social aspect remains a drag on that progress.

Whether or not top-down prioritization occurs—to disseminate, develop, and produce globally accessible knowledge that meets people's basic needs to live in peace, safety, and with dignity, with food, comfort, shelter, and a sense of respect and belonging for their person where they live—grassroots digital communities of practice already can access such life-sustaining knowledge for their communities, no matter how remote, marginalized, or overlooked.

Ultimately, it is not a question of what "we" can do for "them" or vice versa. SAWBO's learning-systems approach recognizes engagement at scales from individual change agents to international coalitions (of mostly global North donors), but the "core" is loosely coupled engagement by communities (of practice). This is not only because digital communities of practice will empirically focus on phenomena to discover knowledge that works and therefore bypass the infrastructural and epistemic ("scientific vs. indigenous") bottlenecks that slow or block both the provision and production of needed local knowledge by others. Nor because they often comprise the only form of knowledge stewardship available for horizontally widening and pragmatizing people's access to the right to knowledge at local scales. Rather, it is because they can "vertically" extend *our* access to empirically grounded knowledge they produce—whether as the production of new (scientifically accurate) language variants, new versions of videos (requested or self-produced using asset libraries), or updating the repository of knowledge that works with additional or new findings.

Whatever the future holds, the vision of this book highlights the indispensable element of partnerships (SDG 17) to reach the SDGs. Its emphasis on loose coupling between parties affords a bidirectionally flexible, ad hoc, sometimes highly informal realization of learning and knowledge support for the local needs of communities that nevertheless connects to and can attain global reach and consequences though an ever-growing online repository of knowledge. Much work and research are still needed to mass-scale these preliminary "answers" to the social problems around empowering everyone's exercise of their right to knowledge, so it reaches the same extent as is now possible for mass-scaling knowledge itself. The hope of this book is not simply that it "lights the candle" of others but sheds light and inspires others to carry forward the quest of pragmatically securing access to the right of knowledge for everyone.

References

Adeyeye, B. A. (2019). African indigenous knowledge and practices and the 2030 sustainable development goal: Exploring its uniqueness for quality knowledge sharing. *Journal of Humanities and Education Development, 1*(4), 147–152. https://doi.org/10.22161/jhed.5.6.7

Agrawal, A. (2002). Indigenous knowledge and the politics of classification. *International Social Science Journal, 54*(173), 287–297.

Ashby, W. R. (1956). *An introduction to cybernetics*. Chapman & Hall.

Bangura, A. K. (2005). Ubuntugogy: An African educational paradigm that transcends pedagogy, andragogy, ergonagy and heutagogy. *Journal of Third World Studies, 22*(2), 13–53.

Bell, D. (1973). *The coming of post-industrial society: A venture in social forecasting*. Basic Books.

Bello-Bravo, J. (2020). Managing biodiversity & divinities: Case study of one twenty-year humanitarian forest restoration project in Benin. *World Development, 126*, 104707. https://doi.org/10.1016/j.worlddev.2019.104707

Bello-Bravo, J. (2023). "New wine in old bottles": Structures of feeling for a new way to see new wood products. *Forests, 14*(3), 524. https://doi.org/10.3390/f14030524

Bello-Bravo, J., Abbott, E. A., Mocumbe, S., & Pittendrigh, B. R. (2020). Identifying and evaluating farmer deviations from steps recommended for hermetic postharvest storage of beans in northern Mozambique. *Journal of Stored Products Research, 87*, 101628. https://doi.org/10.1016/j.jspr.2020.101628

Bello-Bravo, J., Medendorp, J. W., Lutomia, A. N., Reeves, N. P., Tamò, M., & Pittendrigh, B. R. (2022). Dramatically increased accessibility and decreased cost-per-person impacts are needed for scaling IPM in Africa. *Current Opinion in Insect Science, 54*, 100971. https://doi.org/10.1016/j.cois.2022.100971

Brundtland, G., Khalid, M., Agnelli, S., Al-Athel, S., Chidzero, B., Fadika, L., Hauff, V., Lang, I., Shijun, M., & de Botero, M. M. (1987). *Report of the World Commission on Environment and Development: Our Common Future* (O. U. Press, Ed.). World Commission on Environment and Development.

Castells, M. (1997). An introduction to the information age. *City, 2*(7), 6–16. https://doi.org/10.1080/13604819708900050

Dachyar, M., Zagloel, T. Y. M., & Saragih, L. R. (2019). Knowledge growth and development: Internet of things (IoT) research, 2006–2018. *Heliyon, 5*(8), e02264. https://doi.org/10.1016/j.heliyon.2019.e02264

de Oliveira Cintra Farias, F., & Ferreira, H. L. P. (2024). How much for human dignity? A Kant-Maslow compared approach to the basic conditions of life. *Revista Juridica da UFERSA, 8*(15), 136–154.

Delors, J. (1998). *Learning: The treasure within*. UNESCO.

Drucker, P. F. (1969). *The age of discontinuity* (Guidelines to our changing society). Heineman.

Faure, E. (1972). *Learning to be: The world of education today and tomorrow*. UNESCO.

Freire, P. (1970). *Pedagogy of the oppressed* (M. B. Ramos, Trans.). Continuum.

Gadotti, M. (2017). Pedagogía de la Tierra y Cultura de la Sustentabilidad. Foro sobre nuestros retos globales. *Paulo Freire Revista de Pedagogía Crítica, 2*, 61–76. https://doi.org/10.25074/07195532.2.519

Gibson, R. B. (2006). Beyond the pillars: Sustainability assessment as a framework for effective integration of social, economic and ecological considerations in significant decision-making. *Journal of Environmental Assessment Policy and Management, 8*(03), 259–280. https://doi.org/10.1142/S1464333206002517

Golo, B.-W. K., & Yaro, J. A. (2013). Reclaiming stewardship in Ghana: Religion and climate change. *Nature and Culture, 8*(3), 282–300. https://doi.org/10.3167/nc.2013.080304

Hensley, N., & Steer, P. (2019). Ecological formalism; or, love among the ruins. In N. Hensley & P. Steer (Eds.), *Ecological form: System and aesthetics in the age of empire* (pp. 1–17). Fordham University Press.

Horwitz, R. I., Hayes-Conroy, A., Caricchio, R., & Singer, B. H. (2017). From evidence based medicine to medicine based evidence. *The American Journal of Medicine*. https://doi.org/10.1016/j.amjmed.2017.06.012

Kimmerer, R. W. (2012). *Reclaiming the honorable harvest*. https://www.youtube.com/watch?v=Lz1vgfZ3etE&t=1s&ab_channel=TEDxTalks

Knowles, M. S. (1990). *Andragogy in action: Applying modern principles of adult learning*. Jossey-Bass Publishers.

Kumar, A., Kumar, S., Komal, Ramchiary, N., & Singh, P. (2021). Role of traditional ethnobotanical knowledge and indigenous communities in achieving sustainable development goals. *Sustainability, 13*(6), 3062. https://doi.org/10.3390/su13063062

Lakoff, G., & Johnson, M. (1980). *Metaphors we live by*. University of Chicago Press.

Leonard, D. (1995). *Wellsprings of knowledge*. Harvard Business School Press.

Leps, M.-C. (1992). *Apprehending the criminal: The production of deviance in nineteenth-century discourse*. Duke University Press.

Li, B., Liu, M., Liu, J., Zhang, Y., Yang, W., & Xie, L. (2024). Quality assessment of health science-related short videos on TikTok: A scoping review. *International Journal of Medical Informatics, 186*(June), 105426. https://doi.org/10.1016/j.ijmedinf.2024.105426

Locke, J. (1690). *The second treatise on civil government and a letter concerning toleration*. Dover.

Longoni, A., & Cagliano, R. (2018). Sustainable innovativeness and the triple bottom line: The role of organizational time perspective. *Journal of Business Ethics, 151*(4), 1097–1120.

Lutomia, A. N., & Bello-Bravo, J. (2024). Increasing lifelong learning using video animations: The case of an agriculture WhatsApp Group in Kenya. In P. Filippo Gomez (Ed.), *Lifelong learning: Education for the future world* (1st ed., pp. 1-18). IntechOpen. https://doi.org/10.5772/intechopen.114379

Machlup, F. (1972). *The production and distribution of knowledge in the United States*. Princeton University Press.

Mangena, F. (2013). Discerning moral status in the African environment. *Phronimon, 14*(2), 25–44.

Mathevet, R., Bousquet, F., & Raymond, C. M. (2018). The concept of stewardship in sustainability science and conservation biology. *Biological Conservation, 217*, 363–370.

McLeod, S. (2007). Maslow's hierarchy of needs. *Simply Psychology, 1*, 1–18.

Messier, C., Puettmann, K., Chazdon, R., Andersson, K., Angers, V., Brotons, L., Filotas, E., Tittler, R., Parrott, L., & Levin, S. (2015). From management to stewardship: Viewing forests as complex adaptive systems in an uncertain world. *Conservation Letters, 8*(5), 368–377.

Misak, C. (Ed.). (2007). *New pragmatists*. Oxford University Press.

Nemutandani, S. M., Hendricks, S. J., & Mulaudzi, M. F. (2016). Perceptions and experiences of allopathic health practitioners on collaboration with traditional health practitioners in post-apartheid South Africa. *African Journal of Primary Health Care & Family Medicine, 8*(2), 1–8.

Nikolova, B. (2024). The 2030 agenda for sustainable development: Goals and implementation. *International Journal of Sustainability Policy & Practice, 20*(2), 155–180. https://doi.org/10.18848/2325-1166/CGP/v20i02/155-180

Ong, W. J. (1982). *Orality and literacy*. Methuen.

Oyama, S. (2000). *The ontogeny of information: Developmental systems and evolution*. Duke University Press.

Ozuah, P. O. (2016). First, there was pedagogy and then came andragogy. *Einstein: Journal of Biology and Medicine, 21*(2), 83–87.

Panagiotopoulos, P., Vardopoulos, I., Maialetti, M., Ciaschini, C., Koundouri, P., & Salvati, L. (2024). Reimagining sustainable development and economic performance indicators: A human-centric Maslow–Bossel blueprint. *Economies, 12*(12), 338. https://doi.org/10.3390/economies12120338

Petrushenko, Y., & Vorontsova, A. (2020). Knowledge society: The evolution of the concept in the context of achieving sustainable development goals. In W. Leal Filho, A. M. Azul, L. Brandli, P. G. Özuyar, & T. Wall (Eds.), *Quality education* (pp. 511–525). Springer. https://doi.org/10.1007/978-3-319-95870-5_40

Rodríguez-Domenech, M. Á., Bello-Bravo, J., Lutomia, A. N., Medendorp, J. W., & Pittendrigh, B. R. (2023). Digital education and community outreach for sustainable development in Africa. In W. Leal Filho, I. R. Abubakar, I. da Silva, R. Pretorius, & K. Tarabieh (Eds.), *SDGs in Africa and the Middle East region: Implementing the UN sustainable development goals—Regional perspectives* (pp. 1–31). Springer. https://doi.org/10.1007/978-3-030-91260-4_65-1

Rogers, E. M. (1986). Models of knowledge transfer: Critical perspectives. In G. Beal, W. Dissanayake, & S. Konoshima (Eds.), *Knowledge generation, exchange, and utilization*. Westview.

Sandoval-Rivera, J. C. A. (2020). Environmental education and indigenous knowledge: Towards the connection of local wisdom with international agendas in the framework of the sustainable development goals (SDGs). *Diaspora, Indigenous, and Minority Education, 14*(1), 14–24. https://doi.org/10.1080/15595692.2019.1652588

Sattora, E. A., Ganeles, B. C., Pierce, M. E., & Wong, R. (2024). Research on health topics communicated through TikTok: A systematic review of the literature. *Journalism and Media, 5*(3), 1395–1412. https://doi.org/10.3390/journalmedia5030088

SAWBO. (2025). *Scientific animations without borders video library*. https://sawbo-animations.org/

Stone, J. (2021). *Listening to the Lomax archive: The sonic rhetorics of African American folksong in the 1930s*. University of Michigan Press.

Strassman, P. A. (1994). *The politics of information management*. Information Economics Press.

Svensson, G., Ferro, C., Høgevold, N., Padin, C., Varela, J. C. S., & Sarstedt, M. (2018). Framing the triple bottom line approach: Direct and mediation effects between economic, social and environmental elements. *Journal of Cleaner Production, 197*(Part 1), 972–991.

Thayer, H. S. (Ed.). (1982). *Pragmatism, the classic writings: Charles Sanders Peirce, William James, Clarence Irving Lewis, John Dewey, George Herbert Mead*. Hackett Publishing.

Tian, X., Zhang, R., Wang, Z., Kang, X., & Yang, Z. (2024). Exploring community-supported agriculture through Maslow's hierarchy: A systematic review of research themes and trends. *Agriculture, 14*(9), 1568. https://doi.org/10.3390/agriculture14091568

Toffler, A. (1991). *Powershift: Knowledge, wealth, and violence at the edge of the 21st century*. Bantam.

Touraine, A. (1971). *The post-industrial society: Tomorrow's social history*. Random House.

Trosper, R. L. (2007). Indigenous influence on forest management on the Menominee Indian Reservation. *Forest Ecology and Management, 249*(1–2), 134–139. https://doi.org/10.1016/j.foreco.2007.04.037

Umesao, T. (1963). Jōhō sangyō-Ron: Hifu sangyō no jidai no tenkai [Information industry theory: Dawn of the coming era of the ectodermal industry]. *Chūō Kōron, 78*(12), 68–85.

UNESCO. (2005). *Towards knowledge societies (UNESCO world report document SHS.2005/WS/44)*. UNESCO.

von Glasersfeld, E. (1991). Knowing without metaphysics: Aspects of the radical constructivist position. In F. Steier (Ed.), *Research and reflexivity* (pp. 12–29). Sage.

Vygotsky, L. S. (1978). *Mind in society: The development of higher psychological processes*. Harvard University Press.

Waller, D. M., & Reo, N. J. (2018). First stewards: Ecological outcomes of forest and wildlife stewardship by indigenous peoples of Wisconsin, USA. *Ecology and Society, 23*(1), 45. https://doi.org/10.5751/ES-09865-230145

Index

J. Bello-Bravo et al., *The Right to Knowledge*,
https://doi.org/10.1007/978-3-032-03360-4

GPSR Compliance

The European Union's (EU) General Product Safety Regulation (GPSR) is a set of rules that requires consumer products to be safe and our obligations to ensure this.

If you have any concerns about our products, you can contact us on ProductSafety@springernature.com

In case Publisher is established outside the EU, the EU authorized representative is:

Springer Nature Customer Service Center GmbH
Europaplatz 3
69115 Heidelberg, Germany

Batch number: 09301827

Printed by Printforce, the Netherlands